STEAM start

A FULL YEAR OF HANDS ON DISCOVERY

FIRST GRADE STEAM
Units 1 - 5

DESIGN, BUILD, CREATE

Science, Technology, Engineering, Art, Math and Designing in 2D

STEAM start

A FULL YEAR OF HANDS ON DISCOVERY

FIRST GRADE STEAM

DESIGN, BUILD, CREATE Units 1 - 5

Author: Jeannie Ruiz
Editor: Wayne Stripling, Jr.

published by Printing Futures Press

Vancouver, WA

ISBN 978-1-942357-26-1

Photo permission and authors, illustrators, designers, photographers, and editors available for download at the URL http://www.printingfutures.com

STEMWORKS
CERTIFIED
READY-TO-SCALE

STEAMSTART DEVELOPMENT TEAM
STEMWORKS EXEMPLARY PROGRAMS DEVELOPED BY Art of STEM and The International STEM League, Inc.
(Author, Jeannie Ruiz)

Art of STEM developed STEAMStart as well as upper grade level programs that are included in the STEMWorks Database by Change the Equation, a non-partisan, CEO-led initiative to connect and align efforts to improve STEM learning.

RECOMMENDED BY REAL TEACHERS

We are thoroughly enjoying our STEAM lessons. The program is excellent! My students have loved making the Rhombis and getting to build with shapes and tape. It has been a big hit! Thanks for everything!
Alicia Stenard
Mater Christi in Albany School District

"We need more projects like this that carry on for a longer number of weeks. My teachers used scope and sequence suggestions, and the program ran after school for a semester. We could have worked on this a lot longer."
Karen James
Druid Hills Elementary in Charlotte, NC

"I'm excited for our students to get their hands on real STEAM projects. There isn't much out there for Kindergarten and First Grade."
Sandy Mettler
Fort Worth Schools in Dallas, TX

STEAMstart
EDITOR'S COMMENTS

To the Educators using STEAMStart,

First of all, thank you for taking the leap. Project-based learning requires a commitment on the part of the teacher. It can be messy. It can mean more planning and more time spent gathering odds and ends that make a project work. The time and energy you spend developing imaginative responses and ability to problem solve will pay off tenfold.

There is significant evidence in our own research that indicates a need for memorization and rote learning IN ADDITION TO LONG-RANGE PROJECTS and innovative problem-solving. A decent project uses the challenge as a springboard for planning skills-based lessons. A great project urges kids to seek skills and knowledge that you can offer because they want to succeed in solving the problem you've posed. Quick recall of facts is necessary for innovative thinking. Kids don't need to stop and look up simple facts or count on teddy bears when we want them to focus on collecting data and making reasonable decisions.

Projects need outcomes. They need number-driven goals that qualify success in the challenges. Let kids get creative and try things that will fail. Did I say to let them fail? Let them fail. Innovators fail all the time. Some of our greatest inventions are the result of absolute and profound failures. Learning to use the failure as a tool rather than an emotional trial is where students are headed. It takes a few falls before learning to ride a bike. They will make a few bouncy balls on the road to a great slimy wall-walker.

The question arises as to why we chose to call students "designers" rather than "engineers" or "scientists." Designers incorporate a wide variety of communication skills as well as industry expertise. Because communication of ideas among students is integral to the success of any project-based curriculum, the students become Rhombi's designers.

Ten80 celebrates the amazing lessons we learn from you. When you develop a new technique or unique approach to the material, let us hear from you. The resource site grows stronger as teachers share their ideas with each other.

Sincerely,

Jeannie Ruiz
President of Ten80 Elementary

DESIGNERS, WILL YOU HELP ME?

Designers collect, analyze, and synthesize data guided by specifications.

Designers make clear and concise recommendations using drawings,verbal descriptions, and math models.

The Industrial Designers Society of America's definition of industrial design describes design as "the professional service of creating and developing concepts and specifications that optimize the function, value and appearance of products and systems for the mutual benefit of both user and manufacturer."

Rhombi asks her friends to offer assistance in design. Students collect, analyze and make data-based decisions. They must share their concepts verbally, using drawings, and through models (scale and math).

RHOMBI'S ADVENTURES IN 2D STEAM: FIND YOUR RESOURCES

INTRO TO STEAM

APPENDIX

RHOMBI'S ADVENTURES IN 2D
UNITS OVERVIEW

WORLD GOES ROUND (Get moving with "Circle, Circle, Rhombus.") 21

Designers will help Rhombi create a rotating gift for her uncle.

SCIENCE: Chart the moon's shape after reading "Luna and Sol's Big Adventure."

TECHNOLOGY: Take virtual field trips to the moon after reading "Rover, Rover, Come on Over."

ENGINEERING: Create a sand vehicle after reading "Wheels on the Ground."

ART: Use clay circles to create sneaker prints after reading "Quilting in Circles."

MATHEMATICS: Measure circles after reading "Radius and Pi."

SQUARE ONE (Get moving with "Three Rhombis.") 59

Designers will help Rhombi create a tool that proves a square is really a square.

SCIENCE: Use tallies and charts to track squares after reading "Square Foot Harvest."

TECHNOLOGY: Snap pics of shapes for digital presentations after reading "P Is for Parallel."

ENGINEERING: Locate the very useful squares after reading "Shopping for Shapes."

ART: USE measurement to jumpstart quilts after reading "Stories in the Square."

MATHEMATICS: Turn rectangles into squares after reading "Paper Forests."

RHOMBI'S ADVENTURES IN 2D
UNITS OVERVIEW

WE LOVE TRIANGLES (Get moving with "Shape Mimes.") 97

Designers will help Rhombi create a slide for the class pet's habitat.

Science: Test materials for triangle construction after reading "Billy Goat's Bluff.

Technology: Use a virtual geoboard after reading "City of Three Angles."

Engineering: Investigate platform structures after reading "Traveling Triangle."

Art: Play with primary and secondary colors after reading "Fox's Forest."

Mathematics: Reshape the world with polygons after reading "Tri, Tri, Again."

TAKE FIVE (Get moving with "Dot to Dot.") 137

Designers will help Rhombi create new deer crossing signs for the hiking trail.

Science: What's in a pentagon? Take a closer look at the ground inside a regular pentagon.

Technology: Optimize and innovate by designing a tool to improve life in the classroom.

Engineering: Investigate materials under various conditions.

Art: Discover the fun of informational text within a comic strip. Talk about pentagons!

Mathematics: Walk the walls of the pentagon and build its scale model (well, the walls, anyway).

EXTENDING THE ADVENTURE: TINY HOUSE ON WHEELS 174

Designers will take part in an authentic assessment as they create a recreational vehicle.

RHOMBI'S ADVENTURES IN 2D
PROJECT BASED LEARNING

Learning shapes is great. Doing short-term interactive projects that involve shapes is even better. Turning those projects into inquiry-based challenges takes things a step further. Involving students in long-range projects that require multiple steps, depth of knowledge, and true integration of science and math is the goal. The National Science Foundation funded Innovative Technology Experiences for Teachers (ITEST) research that developed a list of criteria for excellence in project based learning.

CRITERIA FOR EXCELLENCE IN PROJECT BASED LEARNING

- Focus on activities that motivate students to develop interest and skills in STEM careers.
- Provide intensive engineering experience of a long duration (academic year and/or summer proportional to the nature of the program).
- Use technology tools and hands-on learning.
- Involve industry mentors, scientists, and engineers.
- Incorporate students' self-regulated learning within specific engineering design experiences.
- Offer innovative, high interest STEM activities.
- Involve participants in the designing and building of artifacts.
- Use scientific integration and/or the engineering design process.
- Engage students in project-based and/or problem-based learning.
- Develop participants' technical skills (as appropriate to the project).
- Make explicit connections between (STEM) academics and project.
- Build in success experiences and opportunities to learn from failure.
- Explore technical careers from technicians to engineers and engineering management.

RHOMBI'S ADVENTURES IN 2D
GEOMETRY IS THE THREAD

The word "geometry" comes from the Greek words for "Earth" and "measure." Geometry was first used to measure and chart the length, area and shape of land surfaces.

GEOMETRY is the single math thread that runs through Common Core from Kindergarten to 12th grade. STEAMStart promotes STEAM and the Art of STEM through an authentic experience in engineering problem solving built on a platform of geometric constructions, analytical geometry, and engineering design. Math is a gateway to success in professional fields: speaking the language of mathematics fluently means that students may define their career paths by choice, not resignation.

~ Jeannie Ruiz

STEAMStart Developer and Director of STEM Ten80 Foundation

THE MATH GENE

Contrary to popular belief, no one is born with a genetic predisposition toward math. There is no "math gene" in humans (Keith Devlin, The Math Gene); mathematics is a learned skill for everyone just like basketball, playing the French Horn, or ballet. Humans have brains which are already wired to enable math acquisition (Stanislas Dehaene, Number Sense 1997). Infants, chimps, dogs, and birds can recognize groups of objects up to three and decide that an object has been taken away or added. That is not "doing math," but it is a qualitative foundation on which to build understanding. Math instruction in school develops the initial qualitative sense of space, time, and quantity into the descriptive -and highly useful- language of mathematics. Children should learn all things in context.

MULTI YEAR FLUIDITY

Geometry as a foundation allows the educator to implement multi-year projects that students enhance year after year. The essence of the project is still fluid, while monthly, weekly, and daily objectives are met for the grade level.

NEXTGEN SCIENCE STANDARDS - STRUCTURE AND FUNCTION

Engineering and Design: Students should understand that "the shape and stability of structures of natural and designed objects are related to their function(s). (K-2-ETS1-2)."

RHOMBI'S ADVENTURES IN 2D
GREAT STEAM TEACHER CHECKLIST

1. **DO MORE THAN IMPART KNOWLEDGE** Children will teach themselves if offered the opportunities to explore and problem solve. Teach by posing questions. Teach by developing curiosity. Teach by guiding learners with interactive experiences. Teach to learn. Learn as much as you teach. Adults often take so much control of the teaching process that children resort to mimicry as a way of pleasing their mentors. Every child learns at a unique pace and with a unique set of circumstances; therefore each child requires a unique path. If educators view learning as a problem-solving process, learning becomes an individual journey.

2. **DISLODGE THE MISCONCEPTIONS** Every child enters school with a unique set of skills and preconceptions about learning. Every child has already formed opinions based on observation, mimicry, and the outcome of challenges (how were successes and failures handled by adults). Some facts that lodge in a child's mind are erroneous. Other ideas are keen observations of the natural and technological world. Facilitate learning. Increase knowledge. Guide inquiry. Open minds to the possibilities. Offer multiple opportunities to learn and approaches to problem solving as a way of addressing misconceptions.

3. **USE WHOLE GROUP SPARINGLY** There are rare teachers with the ability to engage learners as a group. This interaction has a fast-paced Q&A quality and –generally- involves a points system or prizes. The up-side is that learners are involved as a group and learn courteous behavior. The down-side is that educators are rarely able to get individual feedback or offer reinforcement / stimulation. Even technologically based feedback systems are an impersonal way to discover a student's strong and weak areas.

4. **LEARN MORE WITH SMALL GROUPS** Teachers that interact in small groups with students on a regular basis know far in advance of testing exactly what the outcomes will be. Anxiety lessens for teachers and learners. Differentiation becomes possible, and students no longer "tune out" the instructor when the instructor becomes a facilitator. It's hard to ignore the teacher when she's sitting at your table asking questions and challenging assumptions. It's difficult to avoid detection if the investigation requires active participation. For the learner, there is no more "tuning out" or sleep-walking through the day.

5. **ASSESS STUDENTS REGULARLY** There are no surprises when learners take part in authentic and active assessment with reinforcement for some and mental stimulation for others. Great teachers know well ahead of time how students will perform on standardized tests.

RHOMBI'S ADVENTURES IN 2D
ADULT MISCONCEPTIONS

These misconceptions were identified based on in-depth interviews with early childhood teachers about the key issues in early mathematics education as well as researchers' experiences in teaching early childhood students, conducting workshops with early childhood teachers (Ginsburg, Jang, Preston, VanEsselstyn & Appel, 2004; Ginsburg et al., 2006), working with them in early childhood classrooms, and engaging in informal conversations with them. The study also based descriptions of the myths on available research literature (Ginsburg, Lee & Boyd, 2008). The nine misconceptions defined by Joon Sun Lee of Hunter College, The City University of New York and Herbert P. Ginsburg of Teachers College, Columbia University are:

1. Young children are not ready for mathematics education.
2. Mathematics is for some bright kids with mathematics genes.
3. Simple numbers and shapes are enough.
4. Language and literacy are more important than mathematics.
5. Teachers should provide an enriched physical environment, step back, and let the children play.
6. Mathematics should not be taught as stand-alone subject matter.
7. Assessment in mathematics is irrelevant when it comes to young children.
8. Children learn mathematics only by interacting with concrete objects.
9. Computers are inappropriate for the teaching and learning of mathematics.

These misconceptions often interfere with understanding and interpreting the new recommendations on sound early childhood mathematics education, and become subtle (and sometimes overt) obstacles to implementing the new practices in the classrooms (Richardson, 1996).

A WORD FROM TEN80 TO TEACHERS

Become aware of your personal biases. Just because you had poor math instruction does not mean math is hard.

The language of math cannot be taught with the same symbolic approach as the language of reading and writing. Because everyone talks and practices the art of communication, teachers assume that words are the only way to demonstrate proficiency in every field including math and science.

Math and science are performing arts that must be demonstrated in performance. Math and science must use pictures and numbers depicting an actual experiment that is a test of the student's abstract model of some aspect of reality. In this way, students build their own understanding of the patterns that underlie math and science. (Ten80 2002)

RHOMBI'S ADVENTURES IN 2D
DEFINING STEAM

STEAM (science, technology, engineering, arts, mathematics) is just now becoming an accepted acronym among teachers and is seen as a somewhat mysterious –almost magical- fix for all the problems American schools and businesses are facing. Perhaps we should start with what STEAM is not:

- STEAM is not a conglomeration of lessons culled from five subject areas.
- STEAM is not something you can cover in a unit or two every year.
- STEAM is not a 5-day or 10-day curriculum piece.
- STEAM cannot be delivered in the traditional pencil-and-paper method by a sage on the stage standing behind the classroom lectern or seated behind a grand old wooden desk. There are no "Buehler..? Buehler..?" moments in a STEAM environment. *(Reference the movie Ferris Buehler's Day Out)*

SO WHAT IS STEAM?

STEAM is a philosophy of teaching, learning and working. STEAM is more than just the sum of the words in its acronym.

In other words, do real stuff to find real answers to real problems.

STEAM IS PRETTY SIMPLE TO STEAM PROFESSIONALS

What has become a raging debate among educators seems pretty simple to the professionals that actually work in STEAM fields. Academicians will spout research from their PhD thesis and published articles in journals, etc. Ask a scientist, a mathematician, an engineer or a programmer how to prepare kids for a STEAM workplace and they'll tell you to make kids DO something.

SET UP CLASSROOMS FOR DISCOVERY

While activity centers are typical in kindergarten settings, they have all but disappeared by second grade. Bring them back. Small group interaction is optimal for STEAMStart as it is for any STEAM program. Why do you want to use stations or centers in the classroom?

- Simplifies planning - plan over a week rather than day by day...
- Offers opportunities for more cohesive lessons...
- Allows teachers to hold small group sessions daily...
- Encourages teamwork...
- Simplifies classroom management... *
- STEAM increases the ability to draw connections among various concepts.

* Reference — Harry Wong and **The Effective Teacher**
https://www.effectiveteaching.com

RHOMBI'S ADVENTURES IN 2D
DEVELOPING A STEAM MINDSET

While STEAM (Science, Technology, Engineering, Arts, Mathematics combined) is a new buzz word in the field of Education, it was already a well established practice before Mozart gave his first concert as a child prodigy (Salzburg, Austria, 1761).

Learning to incorporate math modeling, grounding the symbolic nature of numbers in reality, and turning the language of mathematics into an integral part of ALL subjects are the true keys to successful STEAM classrooms. Developing confidence in the understanding of size, scale, number, place value, data collection, data analysis and real problem-solving reasonableness is necessary for the creation of a true STEAM culture.

Start each day with a measurement and estimation challenge. As students interact with blocks, toys, manipulatives and other classroom objects, connections should be drawn to their counterparts in the real world.

Ex: Oh, Thomas, that is a 3ft tall tower you've built. Did you know that the Tower of London is 89ft tall? You could tell us a really great story about that tower. Renoir painted towers. Let's look at some of his online. What shapes do you see in the images? Maybe you could paint a Renoir style tower to illustrate your story about the scale model 3ft tower you've designed. If we had more blocks, what else would you add to the tower? A drawbridge? Wow, let's look at the design of a drawbridge's pulley system.

Expecting that level of interaction is realistic when learning centers are guided by the teacher. If all students will be using blocks, you can bet they'll build towers. Anticipating student interests allows the teacher to prepare for verbal interactions that lead to STEAM experiences.

Project-based learning dictates that teachers take the long view in planning. Begin with what you need to accomplish in order to meet some challenge (like addressing Common Core and Next Generation standards). What project will allow natural extensions to your subject matter but also draw kids in as far as interest level? Making the project tie in to real life is also advisable. Do you like to quilt? Incorporate that into your classroom as a tie to measurement, geometry, patterns, history, writing, etc. The project might be designing and making quilts for local shelters. Get creative. In the case of STEAMStart, we've given you the jumpstart. Now, you get to extend the project to other aspects of learning in your curriculum.

RHOMBI'S ADVENTURES IN 2D
CONFIDENT STEAM STUDENTS

Children View Themselves as Scientists in the Process of Learning.

1. They look forward to doing science.
2. They demonstrate a desire to learn more.
3. They seek to collaborate and work cooperatively with their peers.
4. They are confident in doing science;
   they demonstrate a willingness to modify ideas, take risks, and display healthy skepticism.

Children Accept an "Invitation to Learn" and Readily Engage in The Exploration Process.

1. Children exhibit curiosity and ponder observations.
2. They move around selecting and using the materials they need.
3. They take the opportunity and the time to "try out" their own ideas.

Children Plan and Carry Out Investigations.

1. Children design a way to try out their ideas, not expecting to be told what to do.
2. They plan ways to verify, extend or discard ideas.
3. They carry out investigations by: handling materials, observing, measuring, and recording data.

Children Communicate Using a Variety of Methods.

1. Children express ideas in a variety of ways: journals, reporting out, drawing, graphing, charting, etc.
2. They listen, speak and write about science with parents, teachers and peers.
3. They use the language of the processes of science.
4. They communicate their level of understanding of concepts that they have developed to date.
   Children Propose Explanations and Solutions and Build a Store of Concepts.

Children offer explanations from a "store" of previous knowledge.

1. They use investigations to satisfy their own questions.
2. They sort out information and decide what is important.
3. They are willing to revise explanations as they gain new knowledge.

Children Raise Questions.

1. Children ask questions (verbally or through actions).
2. They use questions to lead them to investigations that generate further questions or ideas.
3. Children value and enjoy asking questions as an important part of science.

Children Use Observation.

1. Children observe, as opposed to just looking.
2. They see details, they detect sequences and events; they notice change, similarities and differences, etc.
3. They make connections to previously held ideas.

Children Critique Their Science Practices.

1. They use indicators to assess their own work
2. They report their strengths and weaknesses.
3. They reflect with their peers.

"Inquiry Based Science: What Does It Look Like?" Connect Magazine (published by Synergy Learning), March-April 1995, p. 13.

According to corestandards.org, "The Common Core State Standards provide a consistent, clear understanding of what students are expected to learn, so teachers and parents know what they need to do to help them. The standards are designed to be robust and relevant to the real world, reflecting the knowledge and skills that our young people need for success in college and careers. With American students fully prepared for the future, our communities will be best positioned to compete successfully in the global economy."

Support Common Core Standards and Learning Goals

STEAM PROJECT STATION

Support the ongoing STEAM project with a permanent station and construction materials. Recycled and upcycled materials should be available for use along with tools, paper, pencil and data collection opportunities.

ART, MUSIC, DRAMA

Encourage discussion of STEAM subjects through the arts as well as investigations involving art inventions, art history, genres, musical instruments and sound, etc. Create a "stage" where students give skits, give presentations, teach lessons and make up songs about classroom learning.

BLOCK PLAY

Keep blocks available all the time along with small scale toys at different scales. Add books about world architecture, habitats, and building projects.

NONFICTION LIBRARY

Facilitate the transition from fiction and storybooks to gleaning facts from informational text. Fill your library with fiction AND non-fiction. Encourage students to write informational text and to keep journals for science and math.

STEAM CHARACTER ROLE PLAY

Most kindergarten classrooms include an area for dress-up and pretend. How many first and second graders would still benefit from this type of role-play? Shoot for higher level characterization with lab coats, construction hats, medical scrubs, artist's smocks, etc.

RHOMBI'S ADVENTURES IN 2D
LONG RANGE PROJECTS

If we had a floating house (school, city, state) that could go anywhere on the ocean, where would we want to be?

Instead of being planted in one place, what if we could take our house (school, city, state) with us? Turtles do. No matter where the turtle goes, it's always home.

Will we stay in the same place all the time, or move with currents, weather patterns and seasons?

Build a room, house, school, city, state, or country as a floating island.
Basic elements are squares and triangles which combine (add) to form a solid. Students begin to seek skill in STEAM subjects as a way to solve problems and overcome challenges... as long as the project is interesting enough to keep their attention.

Begin investigating:

- How much milk is delivered to your school everyday?
- Where does it come from?
- What is the source of milk on the floating city-state island?
- How many people can live in a defined habitat? What do they require for survival and comfort?
- How will you move from island to land?
- What happens to structures in various kinds of weather?
- Can we stay out of the way of huge storms over the ocean?
- Do we want to move with the seasons?
- Will it move? How fast do things move?
- If we move, what will fuel the island?
- How much will all that cost, by the way?
- Where is the funding source?

and on... and on... and on...

Since we can't build full-size during the planning phase, we'd better start with scale models. The models will save time, energy, materials, cost and headache. Let's start with the basic information on geometry, polygons and structures.

Introducing STEAMStart.
"Once upon a time, there was Rhombi. She loved shapes..."

RHOMBI'S ADVENTURES IN 2D
USING THE MATERIALS

ENGAGE Each activity uses a read-aloud book to jumpstart the investigation and engage students.

DISCUSS Establish MindBug with an introductory discussion for each investigation. The MindBug is a misconception in student thinking; procedural understanding that worked in one situation but is incorrectly applied to other skills. Like a computer virus, MindBugs spread through a student's system.

Example In Science:
Children often think that air weighs nothing because they see scales measuring zero from an early age.

Example in Math:
Students line the numbers up on the right to add. This works until 5th grade when they attempt to add decimals.

EXPLAIN Each activity includes a brief background for the purpose of introducing the subject matter.

MOVE Get students up and moving to activate that extremely useful muscle memory.

INVESTIGATE

Although each EXPLORE activity and investigation stands alone, together they also build knowledge required to meet the long-range project. In Module 1, students will -ultimately- be challenged to help Rhombi prepare for a trip.

ELA Build skills in communication, writing and speaking.

UNIT 1

WORLD GOES ROUND

WORLD GOES ROUND

Essential Questions

1. How would you describe the moon's shape?
2. What units should be used to measure length?
3. Can you talk about shapes using attributes?
4. How are rotation and revolution different?
5. What are you trying to accomplish in Rhombi's Challenge?

ENDURING UNDERSTANDING

Two dimensional shapes create new shapes.

"Halves" means two equal parts."

Circles and rectangles break into equal parts.

Breaking circles or rectangles into more equal parts means that the parts will be smaller.

Three objects may be put in order from longest to shortest by comparing their lengths.

The length of an object may be discussed using whole numbers.

A smaller object may be used as a measurement tool when measuring longer lengths.

Larger numbers consist of tens and ones.

Organizing information helps in discussion.

Focus on circles and things that move in circles. Learn about rotation and revolution with a focus on the moon. Develop weight-carrying rovers. Dance to the lunar cycle. Explore quilt patterns and create radius art.

Students will apply basic measurement, understanding of number, place value, data collection, and data application in problem solving.

Rhombi wants to create a rotating gift for her uncle. Students design and build a full scale rotating tool, toy, or piece of art. Examples might include pinwheels, pendulums that twist, or waterwheels.

Watch for these extension level words:

astronomer
a person who is skilled in astronomy or who makes observations of celestial phenomena

Galileo
Galileo Galilei (1564–1642) was the first astronomer to use a telescope. He discovered 4 moons orbiting Jupiter and supported Copernicus' theory of a heliocentric universe.

revolution
the action of moving around something in a path that is similar to a circle

rotation
a complete turn around a central point

WORLD GOES ROUND
CURRICULUM CONNECTIONS

Rhombi needs help making a rotating gift for her uncle.

MATH BACKGROUND - Circles

A circle is the set of points in a plane that are equidistant from a given point.

Radius is the distance from center. Diameter is twice the radius. From its center, the circle's angle extends to a full 360 degrees.

TECHNOLOGY CONNECTIONS

Visit the blog for links to excellent STEAM geometry activities and investigations.

- Geometry Games for Kids
- Book Lists
- Links to Apps

HISTORICAL CONNECTIONS

Natural circles like the Sun and Moon exist.

Evidence of wheeled vehicles appears from the mid-4th millennium BC in the Neolithic age.

The first recorded algorithm for Pi was developed in 250 BC by the Greek mathematician Archimedes**.

* Http://en.wikipedia.org/wiki/Circle#History
** Http://en.wikipedia.org/wiki/Pi#History

ART CONNECTIONS

Artists have used circles since earliest recorded history. From mandalas to crop circles, circles appear in all cultural art. Modern artists present the greatest opportunity for class experiences.

Joan Miro was a Spanish painter, sculptor, and ceramicist (1893-1983). Http://joanmiro.com/

Circles in a Circle from 1923 by Wassily Kandinsky, (1866 - 1944) is an oil on canvas.
Http://www.wassilykandinsky.net/

Download free mandala patterns.
Http://www.free-mandala.com/

READING CONNECTIONS

Circle Unbroken
by Margot Theis Raven

So Many Circles, So Many Squares
by Tana Hoban

Pond Circle
by Betsy Franco

Pumpkin Circle: The Story of a Garden
George Levenson

WORLD GOES ROUND
CURRICULUM CONNECTIONS

COMMON CORE CONNECTIONS

ELA/Literacy

I.I.I I use words correctly when I write and speak. I.1.2 I write sentences correctly. L.K.S.C I tell how words are used in real-life. I.1.4 I use strategies to figure out word meanings I.1.4.A I use a whole sentence to understand words in the sentence. W.I.I I write my about a topic and give reasons for my thinking. W.1.2 I write to teach about a topic by giving facts about it. W.1.3 I write to tell an organized story with details. W.1.6 I use a computer or tablet to publish my writing. W.K.8 I use what I know and have read to answer questions. RI.1.4 I ask and answer questions to help me understand new words. RI.1.9 I find facts and attributes that are the same and different in two nonfiction books that teach about the same topic. SL.SI I draw or use other ways to share my ideas. SL.1.6 I speak and share my ideas in complete sentences when needed.

NEXT GEN SCI CONNECTIONS

K-2-ETS1-1. Ask questions, make observations, and gather information about a situation people want to change to define a simple problem that can be solved through the development of a new or improved object or tool.

K-2-ETS1-2. Develop a simple sketch, drawing, or physical model to illustrate how the shape of an object helps it function as needed to solve a given problem.

K-2-ETS1-3. Analyze data from tests of two objects designed to solve the same problem to compare the strengths and weaknesses of how each performs.

K-PS2-1. Plan and conduct an investigation to compare the effects of different strengths or different directions of pushes and pulls on the motion of an object.

K-PS2-2. Analyze data to determine if a design works as intended to change the speed or direction of an object with a push or a pull.*

1-ESS1-1. Use observations of the sun, moon, and stars to describe patterns that can be predicted.

Mathematics

I.G .A.I	I understand and explain what makes shapes unique.
I.G.A.I	I build and draw shapes that have certain parts.
I.G .A.2	I make and draw two-dimensional shapes.
I.G.A.2	I use two-and three-dimensional shapes to create new shapes.
I.MD.C.4	I organize, show and explain number information in a way that makes sense.
I.MD.C.4	I ask and answer questions about number information that is organized.
I.MD.A.2	I talk about the length of an object using whole numbers.
I.NBT.C.4	Sometimes I have to compose a group of higher number when adding 2-digit numbers.
I.NBT.B.2.A	I show that I know what a "ten" is.
K.CC.A.4	I understand how number names go with counting things in the right order.
K.CC.A.4.B	I understand that the last thing I count tells the number of things in a group.

Rhombi Audio Download and Videos Available April 2015

Unit 1: World Goes Round:	Science Activity:	Document the moon's shape over one month.
Unit 1: World Goes Round:	Technology Activity: T	ake a virtual interactive moon or beach tour.
Unit 1: World Goes Round:	Engineering Activity:	Develop a scale model.
Unit 1: World Goes Round:	Art & Music Activity:	Create pattern pendants.
Unit 1: World Goes Round:	Mathematics:	Quilting in Circles
Unit 1: World Goes Round:	ELA Activity:	Read "Rhombi's World Goes Round" aloud.

WORLD GOES ROUND
BUILD STEAM WITH GREAT BOOKS

SCIENCE

So That's How the Moon Changes Shape
by Allan Fowler

Man on the Moon
by Anastasia Suen

TECHNOLOGY

Cars on Mars: Roving the Red Planet
by Alexandra Siyon

Looking Through a Telescope
by Linda Bullock

ENGINEERING

Henry Ford: Big Wheel in the Auto Industry
by Mike Venezia

What Do Wheels Do All Day?
by April Jones Prince

ART & MUSIC

Art & Lots of Dots
by Ginett Alarcon

Eight Hands Round: A Patchwork Alphabet
by Ann Whitford Paul

MATH

When a Line Bends . . . A Shape Begins
by Rhonda Gowler Greene

Shapes in Math, Science and Nature: Squares, Triangles and Circles
by Bill Slavin

ELA

Ladder to the Moon
by Maya Soetoro-Ng

Papa, Please Get the Moon for Me
by Eric Carle

SCIENCE

WORLD GOES ROUND
STEMVESTIGATION: EXPLORE MOON PHASES

READ **Luna and Sol's Big Adventure**

DISCUSS What do students already think they know about the moon? Make a list to identify "Mindbugs."

EXPLAIN The moon rotates and revolves. The Earth's shadow on the moon makes its shape appear to change. The moon reflects light but does not give off its own light. The circular moon we see is actually a sphere, but we see only one side.

MOVE Rotate in place like the Earth or Moon or Sun. Revolve around a rotating student. Try rotating and revolving around the rotating student. Shine a light to simulate the sun. Draw the shapes or snap a photo to show that shadows on the students will also show full, half, crescent, new, crescent, half, full.

INVESTIGATE

Create a chart to mark the moon's shape each day for a month. Assign a quick sketch each day as student homework for the month. Each team of students should be required to demonstrate the Sun's light shining on the Moon. Then each team should be required to demonstrate Earth shadows on the Moon. (Assign one phase of the moon to each team, or have each team demonstrate all 4 major moon phases.)

ELA Review this amazing NASA developed harvest moon video. http://preview.alturl.com/sgvu2

STEMVESTIGATION: MOON PHASES

Draw the moon's shape each night.

MONDAY			
TUESDAY			
WEDNESDAY			
THURSDAY			
FRIDAY			

Luna and Sol's Big Adventure

Explore circles.

Each month, as they moved through the First Quarter of the dance, Luna twirled between Big Sister Earth and Sol,

Big Sister Earth could not see Luna. When that happened, Luna and Sol smiled at one another and Luna twirled faster so Big Sister Earth could start to see her again.

Sol was so big that he cast light on Luna, even if Big Sister Earth was in the way.

Sometimes, Big Sister Earth managed to dance between them so that Luna was in her shadow and Sol could not see Luna.

That did not happen very often.

Luna would giggle and peek out at Sol after a few minutes. Little by little she moved back into Sol's view and continued her dance.

Sol, Luna and Big Sister Earth twirled in their places, rotating in easy circles as they changed places through their dance each month - well, closer to 29 and a half days. Luna's whole life is a dance.

Luna and Sol's Big Adventure

Explore circles.

Every day was an adventure for Luna. She loved to travel and never, ever sat still. Her favorite dance in the world was a twirling rotation. Like a ballerina on a jewelry box, Luna could spin and spin and spin.

Luna's best friend was Sol. Sol lived in the same neighborhood as Luna, but Sol was a lot bigger and older than Luna. Where Luna was cool and calm, Sol was fiery and grand. Where Luna liked to stay near home, Sol traveled in great revolutions around the Milky Way.

Even though they were different, Luna and Sol stayed in touch no matter how much Sol traveled. Sol kept an eye on Luna, warming her with his fiery light. Luna had no light of her own. The side of Luna that faced away from Sol was always cold and dark. But the side that faced him was warmed by Sol's light. Each month - well, closer to 29 and a half days - Luna tuned around one full turn.....)

Luna's whole life was a dance. Big Sister Earth danced around Sol while Luna danced ahead and sometimes behind her, always smiling when Sol poured his light and warmth on her face.

TECHOLOGY

WORLD GOES ROUND
STEMVESTIGATION: ROVERS

READ **Rover, Rover, Come on Over**

DISCUSS Rovers are vehicles handled by drivers on and off site. They are capable of covering difficult terrain without humans if necessary. Rovers were used in moon landings.

EXPLAIN A lunar rover or Moon rover is a space exploration vehicle (rover) designed to move across the surface of the Moon. Mars rovers have landed on the surface of Mars and are busily exploring the planet.

MOVE Place chairs in a mock travel bus formation and take a classroom virtual field trip. Project your tour on the wall or smart interactive board. As you "tour," students lean left, lean right, bounce in their seats, etc.

INVESTIGATE

Visit the Mars Rover site to interact with a virtual surface rover.
http://marsrover.nasa.gov/mission/spacecraft_surface_rover.html

Visit NASA Kids Club online.
Http://www.nasa.gov/audience/forkids/kidsclub/flash/index.html

ELA **Read the winning entries in Poetry Zone's "Space Poems" contest.**
http://preview.alturl.com/sgvu2

Write a poem about the moon!

M

O

O

N

Draw your own rover on the moon.

Rover, Rover, Come on Over

Explore circles.

Two rovers sit still on the moon's surface. The dust on their roofs had not moved in many years. No wind had stirred the little flags attached to the rover antennae. There was no atmosphere on the moon to create weather or wind.

That's Heavy, Man

The LRVs or Lunar Rover Vehicles were very delicate and could never be driven on Earth. They were specially designed for the moon's gravity. Earth's gravity is much higher, making everything on Earth feel heavier than those same objects would feel on Earth.

Apollo Missions 15, 16, 17

The two rovers had been used many times in three lunar visits. Apollo 15, Apollo 16 and Apollo 17 astronauts used the same rovers, parking the vehicles between missions. The Lunar Rover Modules never drove farther than 4.72 miles from the rocket's landing place.

Hammer Time

On the Apollo 17 mission, the hammer in Commander Gene Cernan's space suit pocket accidentally broke one rover's fender. If astronauts hadn't repaired the fender, moon dust would have clogged delicate machinery and equipment.

According to Space.com, the rovers measured about 10 feet long (3 meters), 6 feet wide (nearly 2 meters) and almost 4 feet (1 meter) high.

Rover, Rover, Come on Over

Explore circles.

Duct Tape Designs

On the moon, there are no mechanics to fix a fender, so the astronauts had to use materials they had on hand. The Commander and his team created a new fender using a laminated map and duct tape. You could see the actual EVA* map fender on display at the National Air and Space Museum.

Men on the Moon

The rovers have sat still and silent for more than 40 years. The last visit to the moon launched on December 7, 1972. At the end of that mission, the rovers were present as astronauts placed a permanent plaque on the moon.

"A17-plaque" by NASA - http://images.jsc.nasa.gov/lores/S72-55417.jpghttp://science.ksc.nasa.gov/mirrors/images/images/pao/AS17/10075902.jpg. Licensed under Public domain via Wikimedia Commons - https://commons.wikimedia.org/wiki/File:A17-plaque.JPG#mediaviewer/File:A17-plaque.JPG PUBLIC DOMAIN

ENGINEERING

WORLD GOES ROUND
STEMVESTIGATION: EXPLORE WHEELS

READ **Wheels on the Ground**

DISCUSS What is a wheel? Are wheels all the same size? Are wheels found only in vehicles? Where else might wheels be located? Why couldn't the kids use a skate wheel on the cart?

EXPLAIN Wheels are connected to an object on a rod called the axle like those on bikes, cars, or a set of roller blades. Wheels are also used as doorknobs, clock gears, on pulleys, and in lots of mechanical appliances.

MOVE Drag a student seated in a box across the floor. Now pull students in a rolling chair. Do a relay. One line of students will pull the box without wheels. One line of students pulls the box on wheels (the rolling chair). Compare outcomes.

INVESTIGATE

Student teams will use cut paper, recycled materials, string, thread, tape or other art supplies to make a vehicle that carries pencils, clay or other objects across a tray of sand (simulate snowy ground or moon's surface). The goal is to make the least impression on the sand as possible. Drag, roll, pull, push, move 6 ounces of weight from one end of the sandy tray to the other.

- Fill a large brownie pan with sand (one inch deep at least).
- Supply recycled items and let students use classroom materials.
- Use objects for weight. Students must create the vehicle and container.
- Give a time limit, but let them get truly creative.
- Call time. Teams test their vehicles in the sand pit. Depth of mark left by the vehicle is criteria. Discuss why some vehicles sink or create more depression than others...

ELA Write and illustrate shape poems inside the circles.

WORLD GOES ROUND
STEMVESTIGATION: WHEELS

Round wheels roll across the surface.

Axles connect two wheels.

Wheels may rotate around the axle.

The axle may spin with its wheels.

Wheels on the Ground

Explore circles.

Wheels on the Ground

Explore circles.

Radius and Pi were twins. They looked the same. They even dressed the same, but Radius and Pi were very different people. Radius loved things that were fast. She did not stop to think about what could happen. Pi was a planner. Pi made lists and always made decisions after lots of thinking. Radius and Pi made a good team.

On Saturday, the twins always pulled a cart of organic fresh vegetables and fruits to the local Farmer's Market. The twins grew their own food to sell. They set up a table beside all the other farmers and gardeners. Radius and Pi never missed a market.

This weekend Radius and Pi had a problem. Rain had made muddy puddles on the market path. The cart's skinny tires had sunk 2 inches in the mud. The twins were stuck.

Pi sat down to think. She made notes on a pad and listed all the ways they could solve the this problem. She came up with trucks and sleds.

Radius dove into his project. He liked the idea of a sled. He found a big piece of plastic and piled all the fruits and veggies on the sled. Radius yelled, "Ta Da!" He pointed to his solution and dragged the sled a few feet. It worked but was heavy.

Pi decided to try the truck idea. He found four tires from a toy in the shed. Radius helped Pi prop up the cart. They used a screwdriver and wrench to remove the cart's wheels. Radius held the cart steady as Pi connected each wheel to the cart's axle. The twins lowered the cart to the ground.

Radius gently pulled the handle. He tugged a bit harder. Finally, the cart started moving. The new, wider wheels worked pretty well! Their cart sank only about 1 inch into the mud. The twins moved all the produce to the wagon. The wagon rolled along the path all the way to market.

Radius and Pi sold all the food and had a great day. Best of all, they had worked as a team to solve a tricky problem. Radius and Pi always made a good team.

ART & MUSIC

WORLD GOES ROUND
STEMVESTIGATION: EXPLORE PATTERNS

READ **Quilting with Circles**

DISCUSS How did the author use the patchwork to talk about shapes and the alphabet? Could you make a pattern using only circles? What happens to the quilt edges?

EXPLAIN Circles placed side-by-side will not lay edge to edge. The curve of a circle leaves a gap between each shape. Even tiny circles placed next to each other will never entirely fill the gap. Circles can be overlapped, though, to fill the background space.

MOVE Make people patterns. Give each student a round paper plate. Small teams of students (or the entire class) attempt to make many different patterns using the circles. Kids stand with the plates, using their arms to move the plates into patterns. Snap a quick photo with your phone or camera to review the unique patterns later.

INVESTIGATE

Create round patterned pendants using sneaker prints! Use air dry clay or work with the art teacher to fire the clay. The air drying version is simpler and recommended. Roll a small sphere of clay to a thickness (diameter) of around 1/4 of an inch. Flatten the sphere on waxed paper. Used empty baby food jars or toiler paper rolls as cookie cutters to cut a circle from the clay.

Place the wax paper / flattened clay sphere combination on the floor. Lightly step down with shoe on the clay. Create an imprint. Use a large toothpick or 000000 stick to poke the pendant's hole. Glaze and fire, or dry and decorate.

ELA Create instructions for the pendant. Ask another class to follow your directions, then have a good laugh at their results. Troubleshoot to edit and make directions that work for everyone.

STEMVESTIGATION: PATTERN PENDANTS

Make an instruction sheet for your pendant.
In each circle, write one step.

Quilting in Circles

Explore circles.

Quilting in Circles

Explore circles.

Radius and Pi were visiting Aunt Bee. She had promised to teach the twins how to make a quilt. Aunt Bee's quilts were works of art. Their aunt loved shapes. Today she was helping the twins finish a project. It was full of bold circles sewn by hand.

Aunt Bee had already made the basic quilt. Its bottom layer was squares of old t-shirt fabric. The center layer was fluffy cotton like pillow filling. The top layer was bright fabric, and twins were sewing on this layer. Aunt Bee gave each twin a thimble to keep fingers safe from sharp needles.

Radius and Pi had already used a template to cut circles in lots of sizes. Radius cut circles that measured 1 inch from center, 2 inches from center and 3 inches from center. They had 22 circles in different colors to sew.

Pi drew a pattern of circles that made lots of dots on the fabric. Some of the circles touched edge to edge. Other circles sat one inside another in a quilting pattern called an echo. Each circle was sewn by hand. They hand-sewed their circles one by one.

The planet was rotating away from the Sun when Radius and Pi finished their pro-ject. The Sun sank out of sight, and they could see the moon in the sky. The twins put down their needles. They shook out the beautiful quilt. Circles overlapped and danced on one side. Bright t-shirt squares filled the back. The quilt would always remind them of a great day with Aunt Bee.

MATHEMATICS

WORLD GOES ROUND
STEMVESTIGATION: EXPLORE RADIUS AND CIRCLES

READ **Radius and Pi**

DISCUSS Radius and Pi are interested in circles. The radius of a circle is the length from center to outer edge. The radius of a circle is the same no matter which outer edge you touch while measuring.

EXPLAIN Circles are included in the oval category. All points on the bent line are the same distance from the center of the circle. That distance is called a radius. The entire distance across the circle is the diameter.

MOVE Take a shape walk. As a class, each student should have a blank sheet of paper on which is drawn a circle. Walk around the school and tally all the circles as they are discovered. Extend the lesson: create a map of the school and mark a dot on each circle found. Compare clusters. Where -in the school- are the most circles found? Hint: Keep the walk quiet. Students should make an "okay" sign with fingers to announce each circle's discovery.

INVESTIGATE

Use math to make art. Use the patterns to cut circles with different measures of radius. Place and paste 6 circles on a piece of colored paper (black if possible). Stack another layer of circles of smaller radius but different colors. Repeat. For interest, don't make all the stacks the same. Vary the number of layers. Use different color combinations.

Watch this fantastic video animation. A character walks into Kandinsky's painting to experience its beauty in 2 dimensions. Http://youtu.be/oDN7XEsffBA

Watch this 3D animation of Kandinsky's art. Http://youtu0be/-QcOXmj2VZk

ELA Write and illustrate shape poems around the outside of circles.

WORLD GOES ROUND
STEMVESTIGATION: KANDINSKY CIRCLES

Trace and cut out these circles. Use these as your pattern to trace and cut lots more circles. Use construction paper or paint white paper!

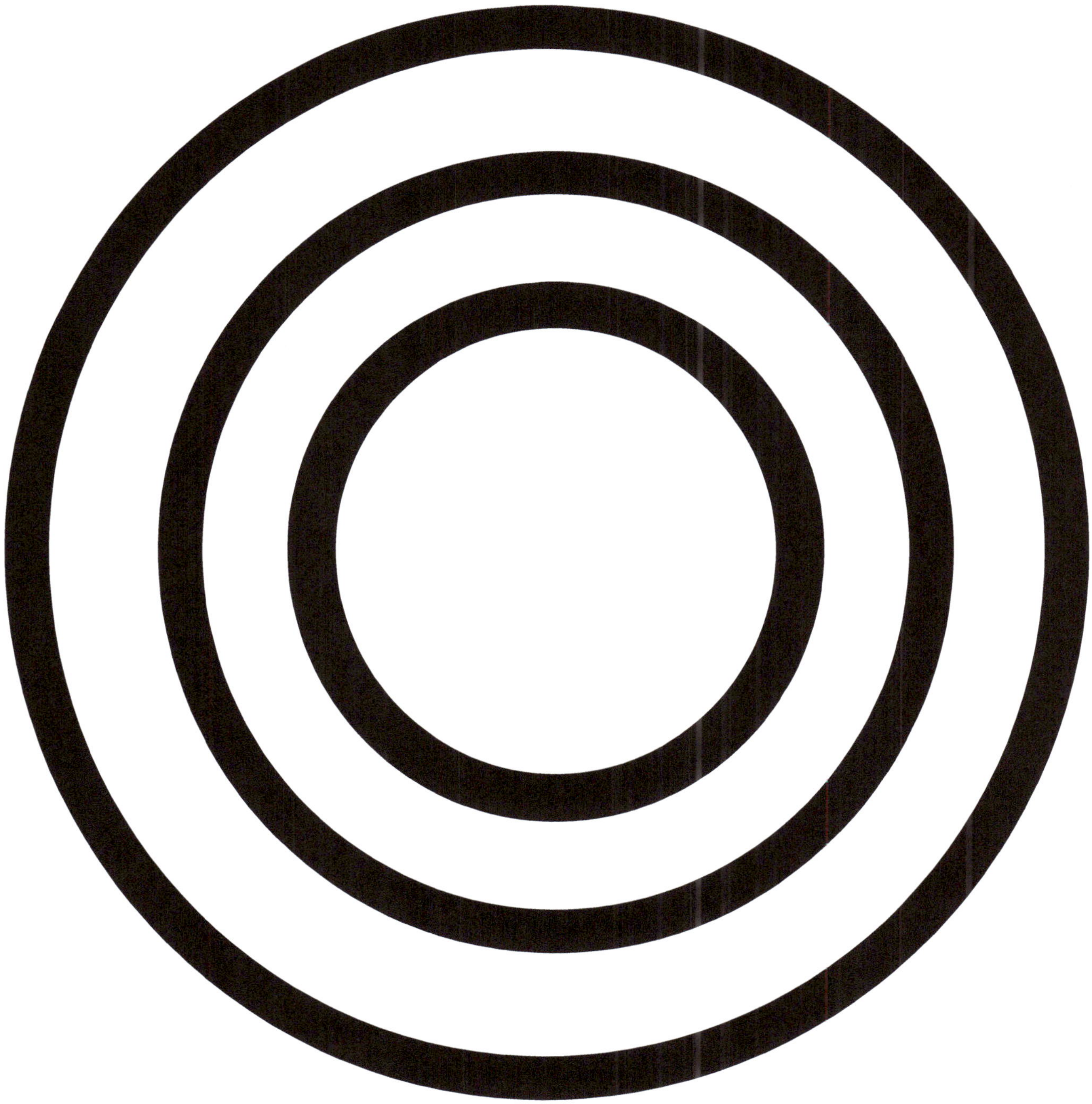

Radius and Pi

Explore circles.

Radius and Pi

Explore circles.

Radius and Pi rolled along the path toward Saturday market. Suddenly, the cart stopped. Radius circled the cart looking for a problem. He leaned down to take a look at the wheel and axle. The wheel had wobbled loose and cracked.

Radius looked around for some way to fix the cart. The twins needed a new wheel. Radius and Pi began looking for round objects that could fit the cart.

The twins found a roller skate wheel and a skateboard wheel in a pile by the curb. Both wheels would be too small. The length from the center hole to the outer part was just too short. Their cart might roll, but it would tilt and spill the apples.

They found some balls, but none had a hole in the center. No hole meant no axle. How would the ball stay on without an axle down the middle?

Radius tried to hold up the no-wheel corner while Pi steered. The two entrepreneurs really wanted to reach the market in time to sell their fruit. They moved too slowly without a wheel. Pi and Radius looked at the bicycles in a shop window. They looked at each other with wide grins and carefully propped the cart on a step.

4 minutes later, Radius emerged with a bicycle wheel minus the tube. It was old and rusted but not bent. The shop owner gave them the wheel. The twins gave her 10 oranges. The wheel was just the right length from center to rim.

Radius and Pi put the wheel on their cart. They zoomed to the market. When the market gate opened, customers were happy to see the twins. Radius and Pi sold all the yummy ripe apples. The twins were proud that they'd solved an engineering problem all on their own.

SQUARE, CIRCLE, TRIANGLE, RHOMBI!

re-imagine the game of Duck, Duck, Goose

MATERIALS

Cut and laminate a variety of polygons.

SETUP

Seat students in a circle. Give each student a polygon. Choose a player to be "Polygon."

INSTRUCT THE PLAYERS

Students sit in a circle on the floor. Each student holds a shape. As "Polygon" walks around the circle, he or she touches each student's head and names the shape that student holds.

ex: circle, circle, triangle, triangle, square, square

"Polygon" chooses a student and says, "RHOMBI."

Continue play as you would with a traditional game of "Duck, Duck, Goose."

UNIT 1 CHALLENGE
WORLD GOES ROUND

WORLD GOES ROUND: PLAY WITH SHAPES

Once upon a time, there was a kid named Rhombi. Rhombi liked shapes and saw them everywhere.

One sunny afternoon, Rhombi and her uncle headed down the street to get some ice cream. Rhombi's Uncle Galileo was an astronomer. He looked for stars, planets and exciting actions in the universe.

Rhombi suddenly turned to her uncle and asked, "Why don't we see the sun all night?"

He smiled and looked up into the night sky. He scratched his short hair, and answered, "That has to do with things going around in circles."

WORLD GOES ROUND: PLAY WITH SHAPES

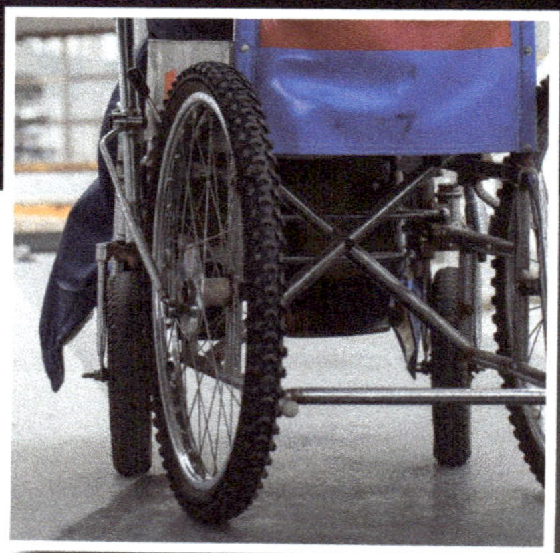

Rhombi's uncle began to spin.

"There is part of your answer," said Uncle Galileo. "I'm the Sun at the center of the Solar System. I spin in place. All the planets spin in place. All the moons spin in place."

Rhombi began to spin. Her uncle said, "Now here is the fun part. You spin and circle my chair."

Rhombi was spinning and walking around his chair. Uncle Galileo started spinning in place at the center of Rhombi's circle. The two got so dizzy they had to stop.

WORLD GOES ROUND: PLAY WITH SHAPES

"The moon spins and revolves around Earth. Earth is always rotating. Part of every day, we are turned away from the Sun," said her uncle. "At the same time, Earth is revolving in a big loop around the Sun. It's like wheels inside wheels inside wheels."

Rhombi thought of all the things she knew that could spin. There were pinwheels and wind-up toys and spinning tops. Uncle Galileo loved all of them.

His birthday was coming, and Rhombi had an idea. She looked out from the pages of her world and asked, "Designers, would you help me create a rotating gift for my uncle?"

RHOMBI'S ADVENTURES IN 2D
TALK ABOUT STEAM

What does the moon look like from week to week?
crescent, half, full, half, crescent

What is an astronomer?
Astronomers study the planets, stars, and events that happen in space.

What does Rhombi's Uncle Galileo do for a living?
astronomer

Where are he and Rhombi headed as they talk?
to get ice cream

How does Uncle Galileo travel?
wheelchair

What is Rhombi's question?
Why don't we see the sun all night, too?

What is rotation?
spin around a center

What is revolving?
Moving in an ellipse / loop / oval / circle around something else...

Can you demonstrate rotation and revolution?
Student stands and spins in place.
Student walks in a wide loop around an object.

TIME TO DESIGN

"Designers, will you create a rotating toy for my uncle?"

LOOK

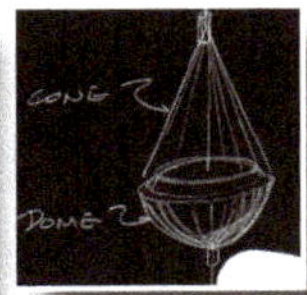

? ? ?

DRAW

FIND

MAKE

RHOMBI'S ADVENTURES IN 2D
CHALLENGE RUBRIC: WORLD GOES ROUND

TASK **Students will use their knowledge of rotation to troubleshoot and build a non-motorized toy that spins.**

	Content	Organization	Presentation
1	• Is well thought out and supports the solution to the challenge or question • Reflects application of critical thinking • Has clear goal that is related to the topic • Is accurate	• Information is clearly focused in an organized and thoughtful manner • Information is constructed in a logical pattern to support the solution	• Format enhances the content
2	• Supports the solution • Has application of critical thinking that is apparent • Has no clear goal • Has some factual errors or inconsistencies (i.e. wheels do not rotate around axle)	• Project has a focus but might stray from it at times (more concerned with form than function) • Information appears to have a pattern, but the pattern is not consistently carried out in the project	• Format does not suit the content
3	• Provides inconsistent information for solution • Has no apparent application of critical thinking • Has no clear goal • Has significant factual errors, misconceptions, or misinterpretations	• Content is unfocused and haphazard • Information does not support the solution to the challenge or question • Information has no apparent pattern	• Presentation appears sloppy and/or unfinished

RHOMBI'S POLYGON ROUND-UP
CIRCLE

ABOUT THE CIRCLE

- In the math world, circle refers to the boundary of the shape. "Disk" is used to refer to the whole shape, including the inside.
- A straight line from the center of a circle to the edge is called the radius.
- A straight line that passes from one side of a circle to the other through the center is called the diameter.
- The distance around the outside of a circle is called the circumference. All points on the edge of a circle are the same distance to the center.
- The value of Pi (π) to 2 decimal places is 3.14, it comes in handy when working out the circumference and area of a circle.
- Circles have a high level of symmetry.

UNIT 2
SQUARE ONE

RHOMBI'S ADVENTURES IN 2D SQUARE ONE

Essential Questions

1. What are the attributes of a regular square?
2. How does technology help you learn about new things?
3. What strategies make counting easier?
4. When would a pattern make a task simpler?
5. What are you trying to accomplish in Rhombi's Challenge?

ENDURING UNDERSTANDING

Two dimensional shapes create new shapes.

"Halves" means two equal parts."

Circles and rectangles break into equal parts.

Breaking circles or rectangles into more equal parts means that the parts will be smaller.

Three objects may be put in order from longest to shortest by comparing their lengths.

The length of an object may be discussed using whole numbers.

A smaller object may be used as a measurement tool when measuring longer lengths.

Larger numbers consist of tens and ones.

Organizing information helps in discussion.

SQUARE ONE

Focus on quadrilaterals, trapezoids, parallelograms, squares and rectangles. Use tallies to "harvest" shapes at lunch. Employ technology in discovering and discussing geometry. Sharpen observational skills with shape walking. Research and recreate famous quilted art. Make your own origami square papers, and learn to fold boxes.

Students will apply basic measurement, understanding of number, place value, data collection, and data application in problem solving.

Rhombi needs to repair the shingles on her playhouse roof. She needs to make each shingle a regular square. Design a tool to check the "squareness" of a shape.

Watch for these extension level words:

builder
In this case, the term "builder" refers to a person that creates or develops a particular thing.

contractor
a manager and coordinator of all the tradespeople involved in a project of many phases

designer
a person who plans the form, look, or workings of something before its being made

engineer
a person who has scientific training and who designs and builds complicated products, machines, systems, or structures : a person who specializes in a branch of engineering

SQUARE ONE
CURRICULUM CONNECTIONS

Rhombi needs a tool to determine square shapes.

MATH BACKGROUND - Circles

Squares are -by definition- regular polygons in that all edges must be the same length. Because they have 4 sides of equal length, squares are regular quadrilaterals. Squares are also rectangles with equal sides and rhombuses with right angles. Square shapes are often used for design and engineering purposes such as town planning.

TECHNOLOGY CONNECTIONS

Build logic and spatial reasoning.

- www.mathplayground.com/tetrasquares.html
- www.mathsisfun.com/games/magic-square-game.html

Mr. R's Math Videos and Songs

- www.youtube.com/user/mathsongs1/videos

HISTORICAL CONNECTIONS

The process of quilting uses a needle and thread to join two or more layers of material. During pioneer days, "paper" quilting became popular. Each piece of cut fabric was basted around paper patterns. Paper was scarce in the west, and women would save letters from home, newspaper clippings, and catalogs to use as patterns. The paper acted as pattern and insulator. Paper found between old quilts is a primary source of information about pioneer life. During the Civil War, slaves used quilts as a means to share and transmit secret messages.

ART CONNECTIONS

Piet Mondrian was born in Amersfoort in the Netherlands, in 1872. Mondrian began his career with traditional art styles, but embraced the concept of "cubism." His most famous painting was "Composition with Red, Yellow and Blue" painted in Paris in 1921.

Cubism was developed by Pablo Picasso and George Braque in the beginning of the 20th century. A typical Cubist painting depicts real people, places or objects, but not from a fixed viewpoint. Instead it will show you many parts of the subject at one time, viewed from different angles, and reconstructed into a composition of planes, forms and colors.

READING CONNECTIONS

Introduce each activity and lesson with a book. Books are listed on the Read about STEAM page. Jot additional lists of books from your favorites here.

1.

2.

3.

4.

5.

SQUARE ONE
CURRICULUM CONNECTIONS

COMMON CORE CONNECTIONS

ELA/Literacy
W.K.7 Participate in shared research and writing projects (e.g., explore books by a favorite author and express opinions). (K-ESS2-1)

Mathematics
MP.2 Reason abstractly and quantitatively. (K-ESS2-1)
MP.4 Model with mathematics. (K-ESS2-1)
K.CC.A Know number names and the count sequence. (K-ESS2-1)
K.MD.A.1 Describe measurable attributes of objects, such as length or weight. Describe several measurable attributes of a single object. (K-ESS2-1)
K.MD.B.3 Classify objects into given categories; count the number of objects in each category and sort the categories by count. (K-ESS2-1)

NEXT GEN SCI CONNECTIONS

ETS1.A: Defining Engineering Problems
A situation that people want to change or create can be approached as a problem to be solved through engineering. Such problems may have many acceptable solutions. (secondary to K-PS2-

1-ESS1-1,1-ESS1-2 Patterns in the natural world can be observed, used to describe phenomena, and used as evidence.

1-ESS1-1, 1-ESS1-2 Make observations (firsthand or from media) to collect data that can be used to make comparisons. Analyzing data in K–2 builds on prior experiences and progresses to collecting, recording, and sharing observations. Use observations (firsthand or from media) to describe patterns in the natural world in order to answer scientific questions.

Mod1: Square One: Science:	**Look for shapes with food.**
Mod1: Square One: Technology:	**Use digital photography to interact with shapes.**
Mod1: Square One: Engineering:	**How can you know that a shape is really square?**
Mod1: Square One: Art & Music:	**Discover the geometry of quilts.**
Mod1: Square One: Mathematics:	**Explore paper squares.**
Mod1: Square One: ELA:	**Read "Square One."**

SQUARE ONE
BUILD STEAM WITH GREAT BOOKS

SCENCE

Pick a Circle, Gather Squares: A Fall Harvest of Shapes
by Felicia Sanzari Senesky

SQUARES
by Sheldrick Ross, Catherine and Bill Slaven

TECHNOLOGY

The Kids' Guide to Digital Photography: How to Shoot, Save, Play with & Print Your Digital Photos
by Jenni Bidner

Looking Through a Telescope
by Linda Bullock

ENGINEERING

Finding Shapes with Sebastian Pig and Friends at the Museum
by Jill Anderson

Shapes : squares
by Sarfatti, Esther

Shapes at School
by Joyce Jeffries

ART & MUSIC

Art & Lots of Dots
by Ginett Alarcon

Dare to Be Square Quilting: A Block-by-Block Guide to Making Patchwork and Quilts
by Boo Davis

MATH

Sam's Sneaker Squares
by Math Matters & Nat Gabriel

My Mother Taught Me Origami
by Manorama Jafa

ELA

Square Bear Meets Round Hound
by Marianne W. Penn

Doodle Dandies: Poems That Take Shape
by J. Patrick Lewis

SCIENCE

SQUARE ONE
STEMVESTIGATIONS: GROWING THINGS

READD **Square Foot Harvest**

DISCUSS What do students already think they know about harvesting? Make a list to identify "mindbugs."

EXPLAIN When a crop has grown and is ready to eat, the farmer must collect it from the fields. This is known as the harvest. Grain crops are collected by huge combine harvesters. Some tree fruits are picked by tall machines. Various crops are harvested all year.

MOVE Pretend to harvest various foods. Will you stretch tall to reach an apple? Should you bend deep to pull carrots or peanuts? Call out a food, and students move through the motions.

INVESTIGATE

"Harvest" the shapes you can find at lunch. What vegetables and fruits are present in student lunch boxes or on their cafeteria trays? Work with students to create a chart for tallying the various foods. Students should offer their educated guesses as to which foods will be represented. After tallying lunch items, compare the guesses with the actual tallies. This is an opportunity to combine counting, data representation, data analysis, geometry and data collection in a simple activity.

ELA Imagine the harvest using five senses. Describe colors, shapes, tastes, sounds, and textures. Students close their eyes and list a descriptive word or phrase per sense.

Tally the vegetables and fruits in your lunch.

Day	Fruits	Vegetables	Dairy	Protein
Monday				
Tuesday				
Wednesday				
Thursday				
Friday				

Square Foot Harvest

Explore garden harvests in science.

Square Foot Harvest

Explore garden harvests in science.

Leaves crunched underfoot as Rhombi headed toward the garden. She kicked a dozen colors into the air just to watch them float back to the ground. Autumn was her favorite part of the year.

Squirrels chittered as they collected acorns from the ground under oak trees more than 70 feet tall. The trees towered above Rhombi like a giant swaying ceiling. She knew these trees must be at least 50 years old if they were dropping so many acorns.

Rhombi and her friends had planted vegetable seeds in the Community Garden in August. Over the past 2 months, the seeds had germinated, pushing out of their hard seed covers. Seedlings began to push through the soil toward the sunlight. Rhombi had spent many hours thinning the rows so each seedling would have enough sun and room to grow. When she saw leaves, Rhombi knew that the radish bulb and roots were reaching down into the soil below. Finally, after almost 40 days, the radishes were ready to eat.

Opening the gate, Rhombi nearly tripped over the huge pumpkins in the first part of the garden. Pumpkins needed lots of room to grow. Each seed had to be planted 5 feet apart. Growing 10 pumpkins took almost 50 feet of garden space. Someone had planted mini-pumpkins that climbed a small trellis at the end of the pumpkin row.

Carrots and radishes and lettuce waited for Rhombi. She pulled 20 carrots and put them into her basket. Next she layered 24 radishes. Finally, Rhombi added 4 heads of lettuce. She picked up her heavy basket and headed home through the crunchy leaves.

She washed and chopped carrots and radishes. She tore lettuce leaves, and added them to a big bowl. When her friends sat down to dinner, Rhombi would proudly serve her delicious autumn harvest salad!

TECHOLOGY

SQUARE ONE
STEMVESTIGATIONS: DIGITAL PRESENTATION

READ **P Is for Parallel**

DISCUSS Digital photography includes all those pictures taken on digital cameras, tablets and mobile phones. How many students take digital photos? Why is George snapping pictures? What is the story's mix-up?

EXPLAIN Digital photography uses light sensitive electronic devices known as an image sensors. These sensors are made up of millions of photoelectric devices that convert light into an electrical signal.

Read more: http://www.practicalphotographytips.com

MOVE Take a mobile phone digital photography walk. Let kids point out objects worthy of a photo. Each time they name an object, the student must also explain what it is using the vocabulary of geometry.

INVESTIGATE

Once back in the classroom, give a slideshow of photos taken and create a PowerPoint presentation of square and quadrilateral shapes found in the school.

ELA "Blog" about the journey around your school and what was found on the investigative hike.

Make a quadrilateral show.

Introduce yourself.

Draw a quadrilateral.

Add photos of quadrilateral shapes.

P is for Parallel

Explore shapes in technology.

P is for Parallel

Explore shapes in technology.

A is for Angle, where two lines meet up...
B is the Base, a 3D shape's bottom or top.
C is for Coordinate, where x and y are.
D is for Decagon, like the drawing of a star.
E is for Ellipse that looks like an oval.
F is for Fractal, every part like the whole.
G is for Google, 1 with 100 zero's.
H is for Hexagon, the honeycomb's hero.
I is for Interior including all the stuffinside .
J is for Junction where things touch or collide.
K is a Kite with four triangles glued together.
L is for Line. Does it goes on forever?
M is for Meter, a unit of measure.
N is for Nonagon, a nine sided treasure.
O is for Octagon. Octa means 8 sides.
P is for Parallel, two lines that never collide.
Q is a Quadrilateral, 4 sides and 4 corners.
R is for Rotation, to turn around a shape's center.
S is for Spheres that are perfectly round.
T is the Triangle that keeps a bridge from falling down.
U is for Undecagon, a 10-sided shape with no order.
V is for Vertex, hanging out at the corner.
W is for Width, a measure from side to side.
X is for X-Axis, a horizontal line.
Y is for Y-Axis, up and down on a graph.
Z is for Zero, no dinner for giraffe?

ENGINEERING

SHOPPING FOR SHAPES
STEMVESTIGATIONS: IDENTIFYING SQUARES

READ **Shopping for Shapes**

DISCUSS From chairs to skyscrapers, every object has a shape. Often, several shapes are hidden in plain sight. Brainstorm a list of locations at which kids see shapes every day.

EXPLAIN Squares have uses in architecture, engineering, and technology. The square must have edges of equal length and 4 right angle corners (vertices).

MOVE Use bodies, arms, legs and books to make shapes as they are called out by "Rhombi". Rhombi says, "Show me a square." Rhombi says, "Make your arms into a circle."

INVESTIGATE

What squares can be found in your school? Create a comprehensive list of square objects and objects that incorporate squares. How will you check for square shape vs. not? Not all things that appear identical are really the same. Hint: remember to look up and down when square hunting. Take a walk to shop for shapes!

ELA Make a diorama of your school or a classroom using cut shapes and gluesticks!

SHOPPING FOR SHAPES
STEMVESTIGATION: IDENTIFYING SQUARES

Take a shape walk. Look for squares.
Draw or describe what you see.

Look up!

Look down!

Look around!

Shopping for Shapes

Explore shapes in engineering.

Shopping for Shapes

Explore shapes in engineering.

Poly and her friend Trap were shopping for groceries. The two were meeting up with Rhombi that afternoon for chess club. It was Poly's turn to bring snacks. She had ten dollars to spend.

Poly pushed a cart up and down the aisles looking for something her friends might like to eat. She dropped a pack of gum in the cart, but it fell through the squares that covered the cart. She placed the gum on the rectangular face of a box of cookies.

Poly wanted to find cheese, crackers, fruit, and cookies. So far, she'd found cookies. They were the round kind with chocolate chips. Trap held up a bag of $3.00 star fruits. They looked like spheres or balls now. When cut, the star fruits would look like little green star shapes on a plate. He also had fat purple grapes that cost $1.00 per bag.

The list included cheese, crackers, fruit and cookies. So far, Poly and Trap had cookies and fruit. The two friends headed toward the dairy section to look for cheese. Wow, there were a lot of cheese shapes. Which one should they choose? There were cylinders of string cheese. Poly pointed to brick shaped rectangular prisms of cheddar. Trap spied a bag of already cut cubes that would be easy to pile on a tray. The $2.00 cheese cubes were added to the cart.

The last item on their snack list was crackers. The cracker aisle was filled with boxes of all shapes and sizes. Oval buttery crackers sat next to 2 inch square saltines and little ½ inch round soup crackers. There were whole wheat rectangles woven like blankets and cheesy 1 inch squares. Some were even shaped like goldfish. Trap picked the $2.50 woven rectangles. They tasted good with cheese.

Their cart was filled with delicious shapes. Star fruits, grapes like spheres, cubes of cheese, and square crackers filled the cart. The total cost was $8.50. They had one dollar and 50 cents left. Poly and Trap took their bags and headed to the chess club. Snack time shaped up to be a yummy break.

ART & MUSIC

SQUARE ONE
STEMVESTIGATIONS: QUILT PATTERNS

READ **Stories in the Square**

DISCUSS Quilts are generally geometric designs created using fabric shapes. Quilt patterns may be simple repeating squares or circles. Patterns may also be complicated.

EXPLAIN Quilts use all shapes as part of images sewn to add interest. Modern quilts have much brighter fabrics than older quilts and often boast more boldly geometric patterns.

MOVE Bring a quilt (or blanket) to play. Students stand around the edge of the square quilt holding its edges with both hands. Pick up the quilt. The teacher drops a balloon on the quilt, and students must work together to pop their balloon up into the air and keep it on the quilt.

INVESTIGATE

Using patterns from a quilting source, use paper to cut-and-paste quilts instead of sewing them. Recreate famous quilt patterns, research modern patterns, and use the resulting geometric artwork as a way to discuss patterns, shapes, measurement, and organization.

ELA Write fictional stories about the paper quilt owners.

Draw your quilt pattern. Use quadrilaterals.

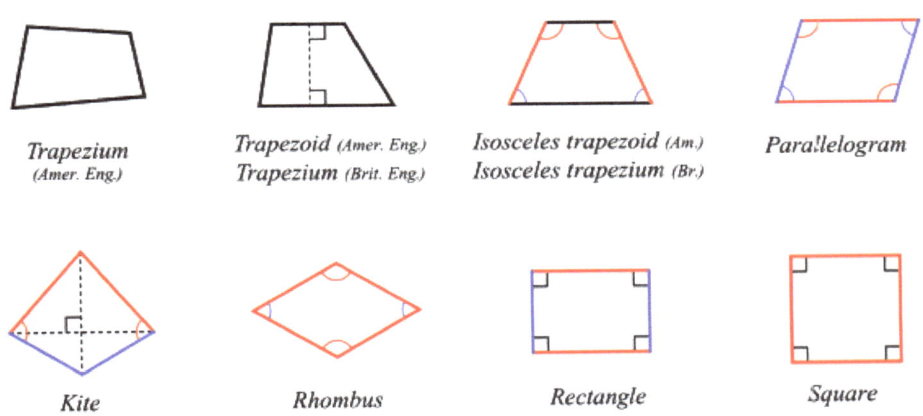

Stories in the Square

Explore square and patterns in art

Quad had a problem. His class was making a quilt for the autumn fundraiser. Each student was responsible for creating a pattern and making one square called a block. All the blocks would be sewn together to make one big quilt. He had no idea what to make.

His floor was littered with crumpled papers. Quad could not choose a pattern. He had drawn and colored and traced until his fingers hurt, but none of the drawings seemed right for his piece of the class quilt. What was Quad going to do? His brain needed a jumpstart.

Quad put down his pencil and headed out to chess practice. He would think more about the quilt block later. At the recreation center, Quad sat down at his usual table. Rhombi sat across the chess board, placing pawns in a row. Behind the 8 pawns, she placed 1 king, 1 queen, 2 rooks, 2 bishops, and 2 knights on the board. The kids took turns moving pieces across the 64 black and white squares. Rhombi almost always played red, while Quad played the white pieces.

As they played, Quad got an idea. He hurried through the match, said a quick goodbye to Rhombi, and hurried home. Grabbing a new piece of paper, Quad folded a square paper in half 3 times. Opening the paper, Quad counted 8 columns. He folded the paper in half 3 times in the other direction. Now he had 64 equal squares.

Quad colored every other square black. He added 8 red pawns, 1 red king, 1 red queen, 2 red rooks, 2 red bishops, and 2 red knights to the black and white board. Quad used his pattern to cut the fabric for his quilt block. He carefully sewed all the pieces together and held the finished quilt block to take a look. It was a really cool pattern.

When the class quilt was pieced together, Quad proudly added his chess board. That week, at chess practice, Quad asked Rhombi for a favor. She agreed with a laugh as Quad set up the Red chess pieces and got ready to play.

Stories in the Square

Explore squares and patterns in art.

MATHEMATICS

SQUARE ONE
STEMVESTIGATIONS: PAPER SQUARES

READ **Paper Forests**

DISCUSS What is origami? Origami is the art of folding paper to look like specific shapes, animals, boxes, and other items.

EXPLAIN Origami papers are shaped like squares, because that shape offers the widest variety of symmetrical folds. Both halves of any origami shape are identical. If the origami bird has a left wing, it will also have a right wing.

MOVE Fold a square! Fold a triangle! Begin with any piece of notebook or construction paper. Fold the top corner down so that A and AA line up exactly. Cut along the resulting line to remove excess paper. Unfold the triangle to reveal a regular square.

INVESTIGATE

Make your own origami square papers. What shapes can you fold from the starting point of a simple square? Experiment with various paper thicknesses. What happens when you fold the same shape from three different types of paper (construction, lined, tissue)? Students should journal about each paper investigated, its attributes, and their observations about its use in origami.

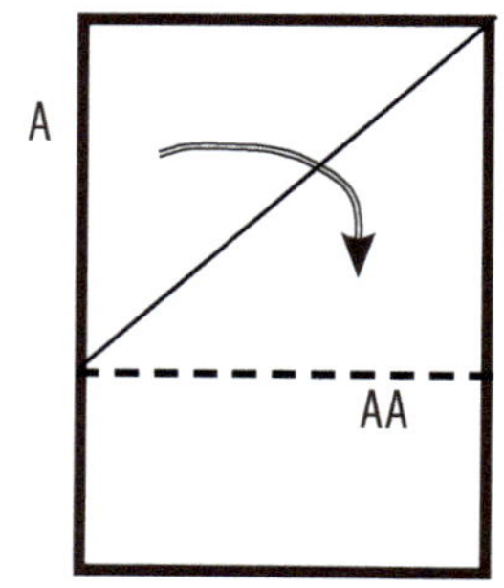

ELA Compile student journal entries into a group guide.

SQUARE ONE
STEMVESTIGATION: PAPER SQUARES

Make your own square papers for origami. Learn to fold a shape.

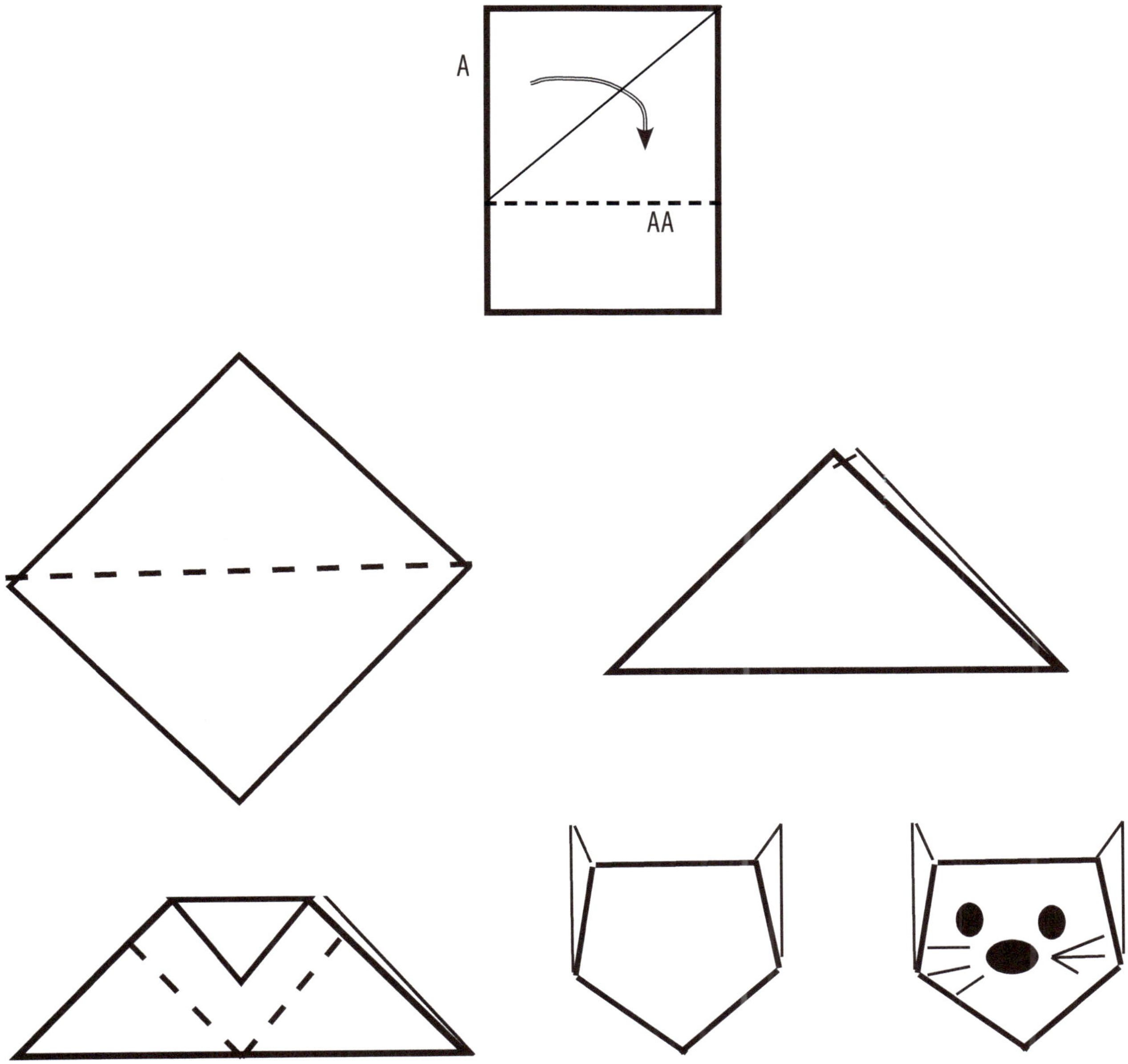

Visit the Internet for patterns and instructions.
http://www.origami-instructions.com/origami-for-kids.html

Paper Forests

Explore origami in mathematics.

Paper Forests

Explore origami in mathematics.

Rhombi's Grandpa folded beautiful paper art. When Rhombi was little, Grandpa could make animals and flowers and boats and castles appear from his hands using nothing but squares of thin paper. Grandpa taught her to fold paper boxes to hold paper jewels. He taught Rhombi to make beautiful princesses with swords that rescued dragons from wicked wizards. Rhombi's favorite project was a mobile hung with 40 folded animals. Grandpa called it his forest in the sky.

Grandpa no longer folded paper, though. His hands were old and gnarled. The fingers couldn't hold delicate paper anymore. Now, instead of the country, Grandpa was living in a city home with lots of other grandmas and grandpas. They all played checkers and had holiday parties. Grandpa seemed happy, but sometimes he talked to Rhombi about the old days. They laughed about a funny fast squirrel that sneaked in through the window and raced around the bedroom before bursting out the front door.

Grandpa remembered pushing Rhombi on the tire swing. He talked about animals that visited from nearby woods. Grandpa especially missed the red fox that used to peer in through the bedroom windows on cold snowy nights. Sometimes they found it sleeping in his shed. Grandpa's new home was in the city. No animals peeked in city windows.

Rhombi planned to visit Grandpa for his birthday on Saturday. Sitting at her desk, Rhombi thought about a gift. She knew he missed his forest and fields, but Rhombi could not bring any live animals into Grandpa's new home. Looking at all the origami papers on her table, Rhombi decided that she could bring some animals to Grandpa after all. She picked up a stack of papers and started to fold.

Saturday was the big day. Rhombi eased her gift through the front door of Grandpa's building. Other grandmas and grandpas and aunties and uncles smiled as she made her way to the dining room. All Grandpa's friends were there with cake and balloons. Everyone oohed and ahhed as Rhombi entered. Grandpa turned to look, and his hands flew to his mouth. He smiled. He grinned. He laughed out loud!

Hanging from Rhombi's hands was a mobile full of forest animals. His squirrel was there in its tree. There were blue birds and rabbits. But, best of all, hanging bright and sassy at the top of Grandpa's new forest in the sky was a beautiful bright red fox. Happy birthday, Grandpa!

THREE RHOMBIS: IDENTIFYING SHAPES

MATERIALS

Objects with faces including circle, triangle, square, rectangle, pentagon
Blindfold

SETUP

Gather shapes OR cut and laminate shapes for each group.

INSTRUCT THE PLAYERS

Students work in pairs or small groups. One student is designated "Rhombi." Rhombi will wear a blindfold and try to name the shape using only touch. Before naming the shape, Rhombi must name the number of edges and the number of vertices. Player earns a point for the complete correct answer. Rotate in a clockwise direction to the next player.

STUDENTS BREAK INTO TWO CLASS TEAMS OR SMALL TEAMS

Each team gets a pile of objects and a blindfold (baggies will keep everything neat). While students play, monitor the discussions. Players should be naming the number of edges and vertices.

UNIT 2

SQUARE ONE

SQUARE ONE:
PLAY WITH SHAPES

SQUARE ONE:
PLAY WITH SHAPES

Once upon a time, there was Rhombi. Rhombi loved shapes. Her friends had helped with a gift for her uncle that rotated in circles! Her own birthday gift had been a telescope. She used it to look at the moon.

Rhombi wondered what she could find on Earth with her telescope. She turned its lens to the Earth. Rhombi saw lots of shapes. She saw circles and all kinds of quadrilaterals. In the yard and around the neighborhood, Rhombi saw a lot of objects with 4 edges and 4 vertices.

Windows on the house looked like squares. Her wagon looked like a square. Tiles on the table and slates on the roof all looked like squares.

SQUARE ONE:
PLAY WITH SHAPES

Rhombi knew that not all quadrilaterals were squares. Some 4 sided shapes that looked like squares were really rectangles or diamonds or kites. She wondered if the roof shingles were really squares? They could be some other kind of quadrilateral.

What had her teacher told the class about squares?

"A square has four equal edges with four corners shaped like the letter L," the teacher had said.

All the shingles looked like the same length, but she would need a ruler to double check. Rhombi noticed a few missing shingles.

SQUARE ONE: PLAY WITH SHAPES

SQUARE ONE:
PLAY WITH SHAPES

Looking at her playhouse through her telescope, Rhombi knew that she should fix the roof before the weather changed. Rain could leak through the gaps between shingles and ruin her playhouse.

Rhombi needed a tool that would help her make sure the shingles were really square. Then she needed a way to make new shingles that were also square.

This was a problem that needed teamwork. Rhombi looked out from the pages of her world and asked, “Builders, can you design a tool like the one this contractor is holding to make sure my shingles are square?”

SQUARE ONE
TALK ABOUT STEAM

What 2D shape do you see in the sky at full moon?
circle / sphere

Make a list of shapes seen from your window.
Answers will vary.

What classroom objects looked like squares?
ceiling and floor tiles

What does Rhombi see through her telescope?
a playhouse roof, windows, table tiles

How do you know if a shape is a quadrilateral?
4 sides / edges

How could you tell a rectangle from a square?
Squares have 4 sides of equal length and 4 "L" shaped angles.
(A square is a rectangle, but a rectangle is not a square.)

What tool does Rhombi need?
identify squares

Can you name some things that have corners like a square?
paper, ceiling and floor tiles

What is a vertice? What does a right angle look like?
A vertice is a corner. A right angle looks like an L.

TIME TO DESIGN

"Designers, will you create a tool to find true squares?"

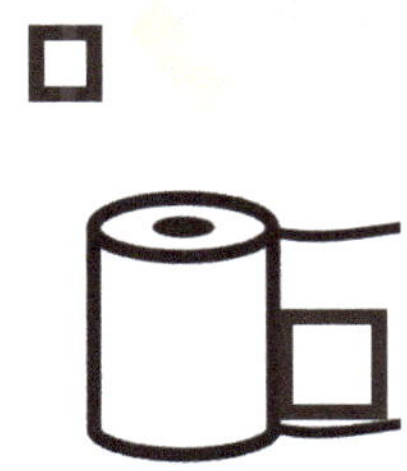

LOOK

DRAW

FIND

MAKE

RHOMBI'S ADVENTURES IN 2D
CHALLENGE RUBRIC: SQUARE ONE

TASK **Students will use their knowledge of squares to design, build, and use a tool for identifying squares.**

	Content	Organization
1	• Is well thought out and supports the solution to the challenge or question • Reflects application of critical thinking • Has clear goal that is related to the topic • Is accurate	• Information is clearly focused in an organized and thoughtful manner • Information is constructed in a logical pattern to support the solution
2	• Supports the solution • Has application of critical thinking that is apparent • Has no clear goal • Has some factual errors or inconsistencies (i.e. wheels do not rotate around axle)	• Project has a focus but might stray from it at times (more concerned with form than function) • Information appears to have a pattern, but the pattern is not consistently carried out in the project
3	• Provides inconsistent information for solution • Has no apparent application of critical thinking • Has no clear goal • Has significant factual errors, misconceptions, or misinterpretations	• Content is unfocused and haphazard • Information does not support the solution to the challenge or question • Information has no apparent pattern

RHOMBI'S POLYGON ROUND-UP
RECTANGLE

ABOUT THE SQUARE

- A square is a polygon with 4 sides of equal length and 4 right angle corners (90 degree corners).
- Because it has 4 sides of equal length, a square is a regular quadrilateral.
- A square is also a rectangle with equal sides and a rhombus with right angles.
- The perimeter of a square is 4 times the length of one side.
- Opposite sides of a square are parallel.
- The internal angles of a square add to 360 degrees.
- A square has 4 lines of reflectional symmetry.

UNIT 3

WE LOVE TRIANGLES

WE LOVE TRIANGLES

Essential Questions

1. How many edges and vertices are present in a regular triangle?
2. Where are triangles found in the building?
3. Why do engineers use triangles in building bridges and other structures?
4. Where would you find triangles in the world?
5. What are you trying to accomplish in Rhombi's Challenge?

Focus on triangles, angles, regular triangles, and corners (vertices). Find apps and games to create triangles on the Internet. Improve Rhombi's cart. Explore primary and secondary colors. Create a triangular world as digital images are remade with triangle overlays.

Students will develop extended measurement skills while deepening understanding of 2 dimensional attributes. Connect with historical backgrounds and introduce tessellations if possible. Watch Vi Hart do some Triangle Doodling. Her videos are as entertaining as they are educational.

Help Rhombi design a fantastic new toy for the class pet using triangles as the strong support.

innovative
using or showing new methods

ENDURING UNDERSTANDING

Two dimensional shapes create new shapes.

The length of an object may be discussed using whole numbers.

Organizing information helps in discussion.

Number information may be organized for the purpose of answering questions.

Explaining number information makes more sense when it is organized.

Shapes may be identified by their attributes.

WE LOVE TRIANGLES
CURRICULUM CONNECTIONS

Rhombi needs to design and build a slide for the bored hamster.

MATH BACKGROUND - Triangles

A triangle is one of the basic shapes in geometry: a polygon with three corners or vertices and three sides or edges which are line segments.

TECHNOLOGY CONNECTIONS

Check out the Math Munch by Vi Hart.
http://mathmunch.org/tag/vi-hart/

Khan Academy's Triangle Doodling by Vi Hart.
http://alturl.com/ntx8p

HISTORICAL CONNECTIONS

Triangles are present in architecture dating to earliest recorded history. Examples include:

- Egypt- Great Pyramids of Giza
- Athens- Parthenon
- Spain- Palais Ideal
- USA - Golden Gate Bridge

Unlike a rectangle, a triangle cannot be deformed without changing the length of one of its sides or breaking one of its joints. In fact, one of the simplest ways to strengthen a rectangle is to add supports that form triangles .

ART CONNECTIONS

9 Pointed Star
This work of art is made up of 16 triangles. Each triangles is equilateral and they rotate to the same degree each time. The triangles are stacked on top of each other. The triangles are used to make a unique star with nine points.

Tilings or tessellations refers to the complete covering of a plane surface by tiles. Pompeii Tiling with equilateral triangles was an example of tessellation.

Go online to create tessellations at this site:
http://gwydir.demon.co.uk/jo/tess/tri.htm

READING CONNECTIONS

Introduce each activity and lesson with a book. Books are listed on the Read about STEAM page. Jot additional lists of books from your favorites here.

1.

2.

3.

WE LOVE TRIANGLES
CURRICULUM CONNECTIONS

COMMON CORE CONNECTIONS

ELA/Literacy
W.K.7 Participate in shared research and writing projects (e.g., explore books by a favorite author and express opinions). (K-ESS2-1)

Mathematics
MP.2 Reason abstractly and quantitatively. (K-ESS2-1)
MP.4 Model with mathematics. (K-ESS2-1)
K.CC.A Know number names and the count sequence. (K-ESS2-1)
K.MD.A.1Describe measurable attributes of objects, such as length or weight. Describe several measurable attributes of a single object. (K-ESS2-1)
K.MD.B.3 Classify objects into given categories; count the number of objects in each category and sort the categories by count. (K-ESS2-1)

NEXT GEN SCI CONNECTIONS

K-2-ETS1-1. Ask questions, make observations, and gather information about a situation people want to change to define a simple problem that can be solved through the development of a new or improved object or tool.

K-2-ETS1-2. Develop a simple sketch, drawing, or physical model to illustrate how the shape of an object helps it function as needed to solve a given problem.

K-2-ETS1-3. Analyze data from tests of two objects designed to solve the same problem to compare the strengths and weaknesses of how each performs.

K-PS2-1. Plan and conduct an investigation to compare the effects of different strengths or different directions of pushes and pulls on the motion of an object.
K-PS2-2. Analyze data to determine if a design solution works as intended to change the speed or direction of an object with a push or a pull.*

MODULE 1: STEAM OPPORTUNITIES FOR "WE LOVE TRIANGLES"

Unit 1: We Love Triangles:	Science:	Find the strongest materials for triangle construction.
Unit 1: We Love Triangles:	Technology:	Use the apps and Internet sites to interact with shapes.
Unit 1: We Love Triangles:	Engineering:	Investigate structures with platforms.
Unit 1: We Love Triangles:	Art & Music:	Use triangles to explore color.
Unit 1: We Love Triangles:	Mathematics:	Reshape the world with triangles.
Unit 1: We Love Triangles:	ELA	Read "We Love Triangles." Students illustrate the story.

WE LOVE TRIANGLES
BUILD STEAM WITH GREAT BOOKS

SCENCE

Showdown at the Food Pyramid
by Rex Barron

Triangles (Shapes in Math, Science and Nature)
by Catherine Sheldrick Ross and Bill Slavin

TECHNOLOGY

The Wing on a Flea
by Ed Emberley
http://www.youtube.com/watch?v=0-YmXsqq_WU

Triangles and Pyramids
by Sally Morgan

"The Polygon Song"
by Steve Saunders
http://youtu.be/TTrbT_SSD0U

ENGINEERING

Triangles Around Town
by Nathan Olson

The Bridge
by Emily Cheney Neville

The Art of Construction
by Mario Salvadori

ART & MUSIC

Triangle for Adaora: An African Book of Shapes
by Ifeoma Onyefulu

Draw a Triangle, Draw Anything!
by Chris Hart

MATH

Pop's Bridge
by Eve Bunting

Mouse Shapes
by Ellen Stoll Walsh

The Greedy Triangle
by Marilyn Burns

ELA

WHAT ARE YOU READING?

SCIENCE

WE LOVE TRIANGLES
STEMVESTIGATIONS: POLYGONS

READ **Billy Goat's Bluff**

DISCUSS Where do you see triangles?

EXPLAIN Polygons may have three, four, five, six, or more sides. All sides must be straight. Triangular polygons are more stable than other polygon shapes. The angles in a triangle cannot change once the triangle is built. The angles are fixed. Because a triangle has three sides and three angles, each angle is fixed by the side opposite to it.

MOVE Use chalk and straight edges to mark regular polygons on the sidewalk. Use chalk and straight edges to mark irregular polygons on the sidewalk.

regular triangle, regular square, regular pentagon, regular hexagon
Irregular triangle, irregular square, irregular pentagon, irregular hexagon

INVESTIGATE

Use straws and clay (or marshmallows) to construct regular polygons in the shapes of regular triangles, regular squares and regular pentagons. Flex each polygon. Which one seems least able to bend or change its corners (angles)? Compare the strength and stability of each shape. Make polygons from many different materials. Use clay, marshmallows, foam, gum and any other connector to prove that the shape leads to the triangle's strength rather than any fixative. Keep careful records, and make sure to use the same amount of material in each test run.

ELA Create a journal entry regarding the "tests" performed on each material. Present findings to the class.

WE LOVE TRIANGLES
STEMVESTIGATIONS: GROWING POLYGONS

Use straws and food to make these polygons.

1. How many polygons can you name without peeking at the answers?
2. Trace each of the polygons on this page.
3. Write the name of each polygon you cannot name without peeking.

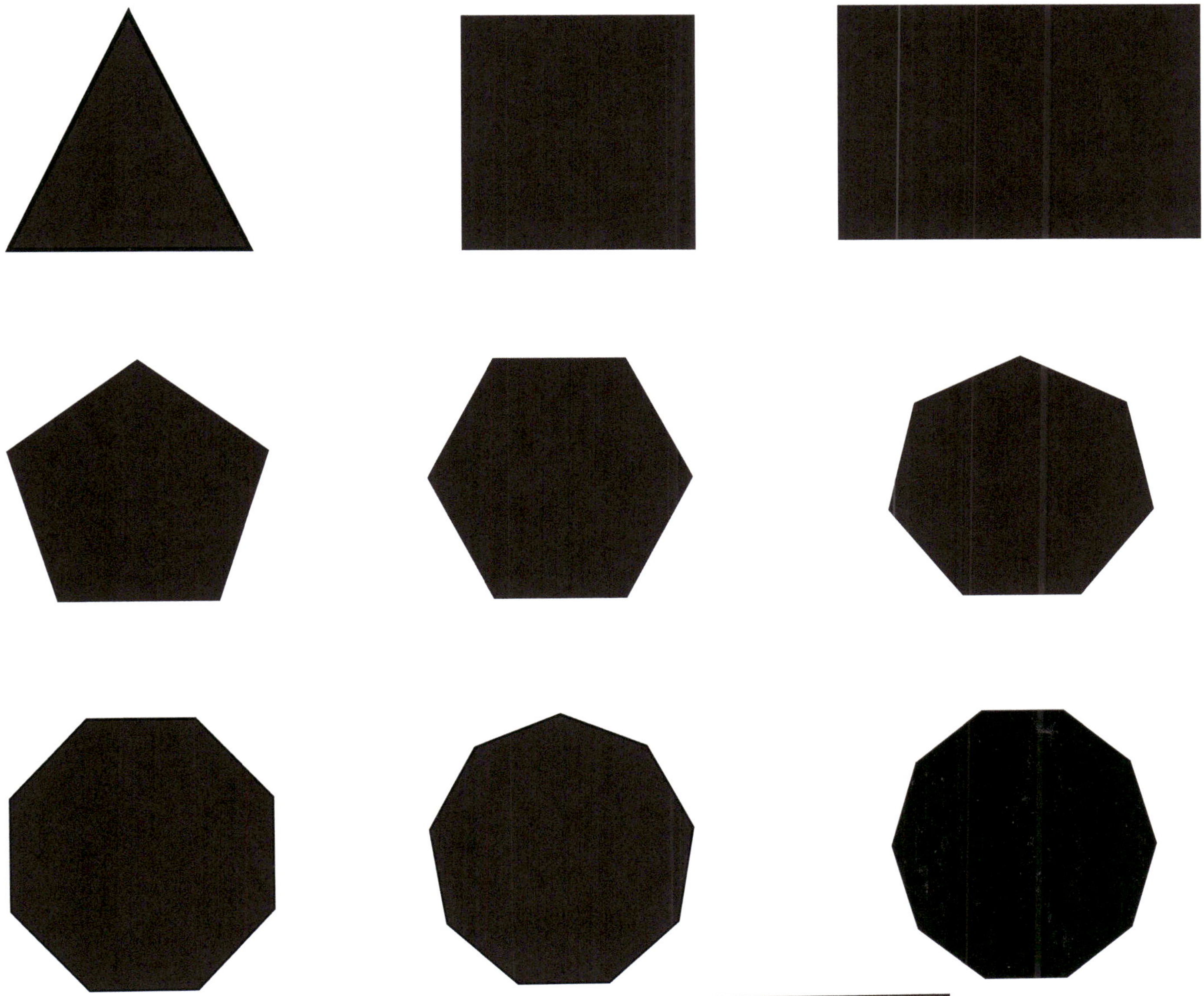

Triangle, square, rectangle, pentagon, hexagon, heptagon, octagon, nonagon, decagon

Rickety Rackety

Explore structures in mathematics.

LEFT COLUMN

MAN MADE THESE BRIDGES

a. trestle b. suspension c. coins stacked

RIGHT COLUMN

NATURE MADE THESE BRIDGES!

a. beaver b. fallen log c. eroded rock

Rickety Rackety

Explore structures in mathematics.

Once upon a time there were three little goats. They were hungry and wanted to eat clover at the top of the triangle hill. To reach the clover, each goat had to cross a rickety rackety bridge.

Youngest goat began to cross the rectangle bridge. He dashed across the 12 feet of bridge as it swayed back and forth under his weight. Rickety rackety rickety rackety went the bridge. The beaver under the bridge shouted at them to go away. "You are too loud, and you're making my bridge go rickety rackety."

Youngest Goat answered, "I hope the square-bottom bridge will hold. My brother is on the way, and he is much bigger than I."

After 30 minutes, Middle Goat began to cross the bridge. Rickety rackety rickety rackety went the bridge. The beaver under the bridge lumbered out and shouted at them to go away. "You are even louder than your brother, and you're making my bridge go rickety rackety."

Middle Goat answered, "I hope the square-bottom bridge will hold. My brother is on the way, and he is much bigger than I."

Along the path came Biggest Goat. He took one look at the bridge and shook his head slowly. His triangular coal black beard waggled back and forth. The biggest goat put one hoof on the bridge. Rickety rackety rickety rackety went the bridge. The angry beaver underneath shouted at them to go away! The goat peered under the bridge and saw that the bridge was held up by a single square. He looked across the stream at his brothers. If he couldn't fix the bridge, the younger kids would be stuck on the hillside. Biggest Goat thought and thought. What was the strongest shape he could remember?

Youngest Goat saw a seed and bleated, "What about a circle?" Biggest Goat shook his head. The bridge would roll away. Middle Goat saw a flower petal and bleated, "What about an oval?"

Biggest Goat shook his head. The bridge would rock back and forth like a rocking horse. He needed a very powerful shape to keep the bridge straight and strong. Finally, Biggest Goat looked up at the mountain that held the delicious clover. What shape is a mountain? It's a triangle. Worn by weather and wind, the mountain stands strong for thousands of years. Biggest Goat asked the beaver to help him with the bridge. The beaver grumbled but made lots of triangles to hold up the bridge. Finally, Biggest Goat began to cross the bridge. The happy beaver waddled back under his very strong bridge. He didn't even notice when the three brothers headed back across the bridge at sunset.

TECHOLOGY

WE LOVE TRIANGLES
STEMVESTIGATIONS: MAPS AND ANGLES

READ **City of Three Angles**

DISCUSS Angles are the measurements used to define shapes in geometry. Triangles are described by their angles and number of equal edges.

EXPLAIN Triangles are grouped in three categories: acute angles are smaller than an "L" shaped angle. Right angles are "L" shaped angles. Obtuse angles are more open than "L" shaped angles.

MOVE Use arms to form acute, right and obtuse angles. Make up a class dance for triangles!

INVESTIGATE

Visit NCTM to play with triangles on the interactive GeoBoard.
http://www.nctm.org/standards/content.aspx?id=25007

Visit MathCats to interact with Polygons.
http://www.mathcats.com/explore/polygons.html

Make symmetrical pictures at Kali.
http://www.scienceu.com/geometry/handson/kali/

Draw the steps for your class triangle dance!

Step One

Step Two

Step Three

City of Three Angles

Explore maps in technology.

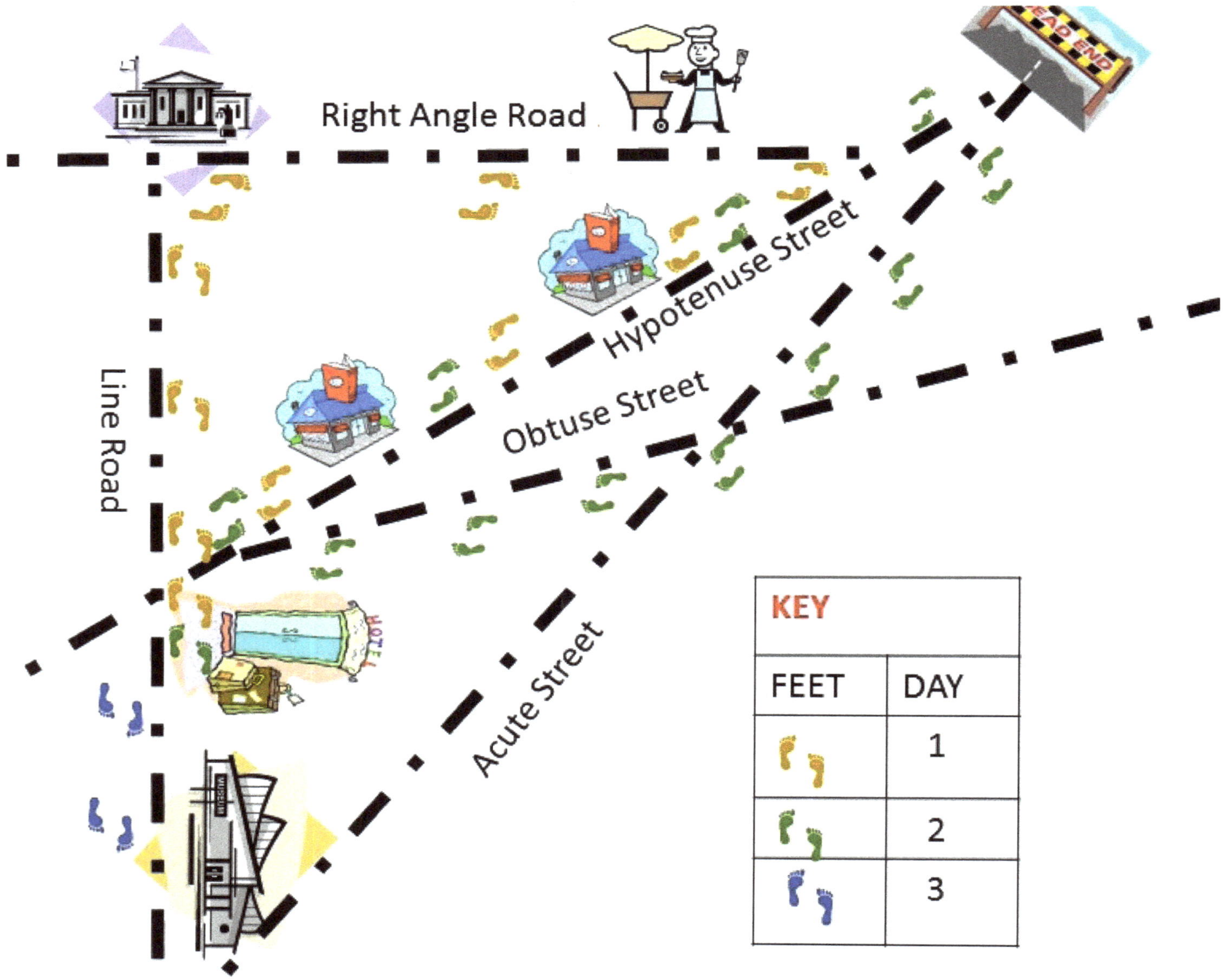

City of Three Angles

Explore maps in technology.

Every street in Three Angles was named after the shape of its corner. That made walking through town easy if you knew your shapes. Quad wanted to find Polygon Museum. He was new to the city, but he was pretty sure he could get around. Quad didn't bother taking a map.

Quad walked North along Line Street to the town square. He turned right at the square and walked East along Right Angle Road. Quad bought a cool dog collar at the Pet Palace. At the farthest end of Right Angle Road, Quad took a sharp right turn and headed South along Hypotenuse Street. He found two great bookstores along the way. Finally, Quad ended up right back where he'd started! He had not found the museum. He decided to try again on Friday.

The next morning, Quad looked for the Museum again. This time, he started on Hypotenuse Street. He headed North on Hypotenuse until the street came to a dead-end. Quad decided to turn right onto Acute Avenue. He was sure to find the museum at the end of Acute Avenue. Quad reached the corner and took a slight right onto Obtuse Place. He walked and walked until he saw the sign for his hotel at the end of Obtuse Place just before Hypotenuse.

Good grief! The sun was setting. He'd have to find the museum on Saturday. This time, he'd save time and use a map.

On Saturday morning, Quad stood outside his hotel. The young tourist looked at his map. He touched the name of the Polygon Museum on the map. Slowly, Quad looked left. The museum was beside his hotel. Had Quad turned left out of the hotel, he would have seen the Museum right away. Next time, Quad would definitely use a map.

ENGINEERING

WE LOVE TRIANGLES
STEMVESTIGATIONS: STRUCTURES

READ **Traveling Triangle**

DISCUSS "The Art of Construction" draws on examples from every period in history to explain the principles of building. Why do buildings and bridges stand up? How do they withstand the tremendous forces of wind, earthquakes and gravity? Talk about which blocks and toys stand up to student destruction.

EXPLAIN Triangular polygons are more stable than other polygon shapes. The angles in a triangle can not change once the triangle is built. The angles are fixed. Because a triangle has three sides and three angles, each angle is fixed by the side opposite to it.

MOVE Look for examples of triangles making structures strong. Students play a version of "musical chairs" using triangles in objects. Play music while students look for examples. Stop music. Anyone left without an example must sit down...

INVESTIGATE

Challenge students to build a platform using 4 blocks and cardboard. How much weight may be balanced on the platform before it topples?

Challenge students to build a cardboard platform held up by 3 blocks in a triangular formation. How much weight may be balanced on the platform before it topples?

ELA Create a list of objects that include platforms. i.e. attic floor, porch, table, chair, etc.

Make block tables to hold books and toys.

Make a table.	**Count blocks.**
1 Block and Cardboard	Blocks
2 Blocks and Cardboard	Blocks
3 Blocks and Cardboard	Blocks
4 Blocks and Cardboard	Blocks
5 Blocks and Cardboard	Blocks

Traveling Triangles

Investigate structures.

Traveling Triangles

Investigate structures.

Triangle, Triangle,
where have you been?
I've soared in the sky
blown East and West by the wind.

Triangle, Triangle,
what have you seen?
I've watched the city
from the top of a Ferris wheel.

Triangle, Triangle,
what have you done?
I've held the wood steady
for builders in the hot summer sun.

Triangle, Triangle,
where have you gone?
I've biked along city streets
in the cool quiet dawn.

Triangle, Triangle,
what should we play?
Let's climb on the jungle gym
at the end of the day.

ART & MUSIC

WE LOVE TRIANGLES
STEMVESTIGATIONS: TRIANGLES IN ART

READ **Fox's Forest**

DISCUSS How many objects can be drawn using triangles? Where do you see triangles in faces? Birds? Lizards?

EXPLAIN The triangle does not exist as a regular polygon on animals. Triangles are something we've created, but we can recognize the essential shape using our imaginations.

MOVE Laminate a triangle (and other shapes). One team of students should affix the shape to an object in the classroom that contains a triangle. Other teams attempt to locate the shape first. Play progresses as teams take turns placing the triangle (and other shapes) until reaching 180 points. 180 is the total sum of all three angles inside a triangle.

INVESTIGATE

There are three primary paint colors (blue, red, yellow), the three pigments that form all art. Photocopy a regular equilateral triangle for each student. Paint one edge blue. Paint one edge red. Paint one edge yellow. Investigate the blending of primary colors. Using a paintbrush, gently pull some of each color toward the triangle's center. What colors result from blending all three? These colors are called secondary colors.

ELA Make lists of objects in each color group. Describe emotions that might correspond to the various colors.

STEMVESTIGATION: TRIANGLES IN ART

Paint the edges of your triangle in primary colors.

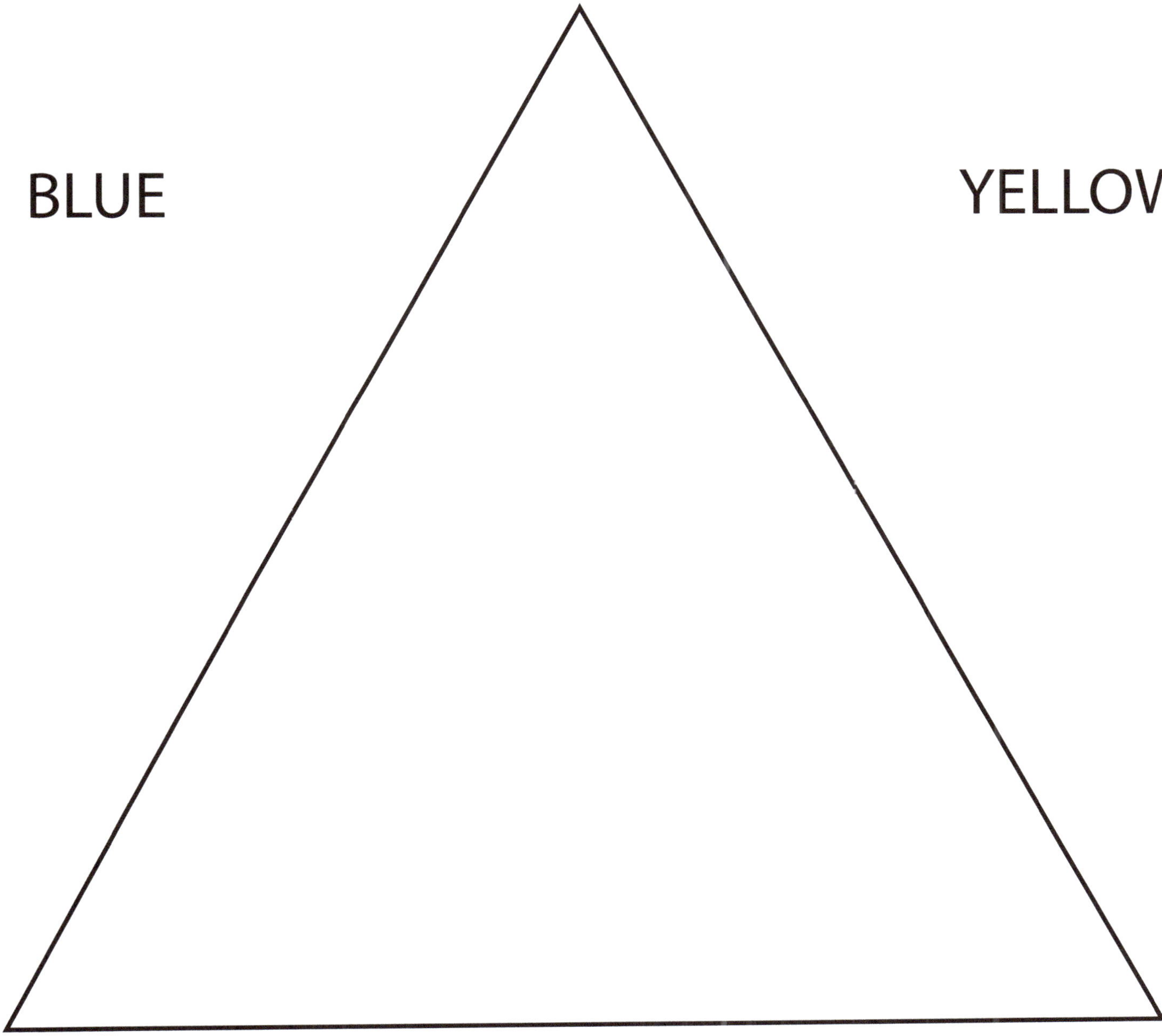

Fox's Forest

Explore environments in art.

Fox's Forest

Explore environments in art.

Fox trotted happily on the warm summer path. Sun beamed down through leaves to make moving patterns on the worn dirt track. Fox's head swung from side to side looking for yummy treats. He also scanned the woods for anything that might want to eat him!

Bushy tail wagging, Fox stopped to nose some flowers. Soon these green leaves would part to reveal little red berries. Delicious and sweet, Fox loved these berries in the straw. Fox caught another smell in the straw. It was the ripe odor of rotting log. Logs were good. They often held tasty crunchy bugs.

Fox lifted his little pointed face to the wind. Where was the smelly log? He faced West toward the setting sun. Sniff, sniff. No logs there. He faced East toward the far edge of the woods. No log smell rose from that direction. Fox turned North to sniff. There. The stinky log was that way!

Nose to the ground, Fox left the path and headed into the green woods. Shadows played across Fox's rust-colored body. Sniff, sniff. He followed the air molecules that blew in the breeze from that stinky log. The molecules were pulled in through Fox's nose and mouth. His brain made sense of the smells, and Fox turned a little to the West. Finally, behind a tall rock, Fox found a log so rotten it made his mouth water. Using tiny paws and claws, Fox pulled at the wet bark. Beetles and mealworms tried to scatter. Fox caught them and crunched his way through an afternoon snack. Delicious!

Pointy little face covered in soil and bits of rotten tree, Fox turned back toward the path. South and a little East brought Fox right back to strawberry leaves beside the warm summer path. It wasn't quite so warm now that the Earth was turning away from the Sun. Shadows crept closer as Fox found his way back to the den. With a full belly, Fox curled up with his tail over his nose and fell asleep in the safety of his dark, dry hole.

MATHEMATICS

WE LOVE TRIANGLES
STEMVESTIGATIONS: ANGLES

READD Tri, Tri Again

DISCUSS Bird feeders are a constant source of joy and annoyance for avid bird watchers. Anti-squirrel devices are required to keep rodents out of the bird feed. Triangles are used in this story, but many other shapes are used in these ingenious traps.

EXPLAIN Geometric shapes used in mathematics have very specific rules to keep all mathematicians on the same page. Triangles must have corners (angles) with measurements that add up to exactly 180 degrees. Regular triangles must have edges / sides that are all the same length.

MOVE Take a digital shape walk. Snap photos of various scenes around the school that look interesting shape-wise.

INVESTIGATE

Using your digital images of the school, choose a few for students to use. In a presentation software program, have students place shapes over parts of the picture. For example, place a triangle over the angled rooftop. As each group takes a turn with the image, they will add their own shape replacements. Eventually, your class will create a fully re-shaped image!

Free versions to download:

Empressr
http://www.empressr.com/

SlideRocket
http://www.sliderocket.com/

Prezi
http://prezi.com/

WE LOVE TRIANGLES
STEMVESTIGATION: ANGLES

Re-shape your world. Use shapes to cover the image.

Tri, Tri Again

Explore triangles in mathematics.

Tri, Tri Again

Explore triangles in mathematics.

Poly really needed to fix her bird feeder's roof. The feeder hung from a branch outside her window. All kinds of feathered friends relied on the feeder for their winter meals. Sadly, there were a bunch of furry friends that also wanted to eat at the hanging seed source. Long fluffy tails knocked the feeder to the ground over and over again. Poly was afraid that the feeder would break in one of its squirrelly falls.

Poly had tried lots of different ways to trick the squirrels. She wanted to keep them away from the feeder, but the mammals wanted all the free seeds. Squirrels had nuts of their own, but the bird feeder was such an easy food source. Poly had designed a round tipsy cover that should have made the squirrels wobble and fall to the ground. They learned how to climb around it in less than a week. Poly had turned a mesh scarf into a wrap that squirrels couldn't get their hands through. They'd just untied and dropped the rectangular scarf to the ground. Poly had even covered the bird feeder hanger with drops of slippery oil. The greedy beasts just slid down it like a fireman's pole. They even seemed to have fun!

Poly had thought she was out of ideas. Round plates hadn't worked. Rectangular scarves hadn't worked. Spheres of oil hadn't worked. That afternoon, on her class trip to the city museum, Poly had seen a tall pointed roof. The roof covered a bell, and the guide told her class that the roof kept pigeons from landing. Bingo! If it kept pigeons away, the roof should keep squirrels away.

Poly cut out 3 triangles of wood. She covered them with waterproof material used for raincoats and connected all three triangles at their long edges. She formed a tall pointed object and attached it to the bird feeder. Her brand new triangular roof was nearly 3 feet tall and very steep. She hung the modified feeder and sat down to wait.

Birds flew in and out to get their seed meals. Finally, two fat little squirrels came to sit on the branch above the feeder. They tilted little heads to one side and the other side looking confused. Finally, one brave little friend leaped to the roof. She slid right past the seed and landed on the ground. That was one grumpy squirrel. Her rodent friends each tried the feeder, but all failed. Poly had learned her lesson.

If at first you don't succeed with circles and rectangles, tri, tri again.

SHAPE MIMES

MATERIALS

Pictures of shapes or shape names written on paper

SETUP

Break students into teams of 3 or 4 students. Place shape pictures or names in a bag or hat. Create a points tally chart on the board.

INSTRUCT THE PLAYERS

Teams will compete against one another for points OR adds points team-by-team to the whole class goal. Choose team names. One person from the first team chooses a shape picture / name from the bag. The team members must convey to students which shape they've picked BUT MAY NOT SPEAK AT ALL.

- Point to an example in the room.
- Make the shape with your arms.
- Form the shape with your bodies (3 kids at the vertices of a triangle, etc.).
- Trace the shape in the air.

UNIT 3 CHALLENGE

WE LOVE TRIANGLES

WE LOVE TRIANGLES: EXPLORE SHAPES

Once upon a time there was Rhombi. Rhombi loved shapes. Circles were cool (there were circles in the word "cool"). Her innovative helpers had designed a tool to find squares. That had been a fun project.

Rhombi sat in the park. She looked around the playground and played her own game of "I SPY" with shapes.

"I spy with my little eye a square tile."
"I spy with my little eye 80 circles."

80 circles were linked to make a chain that held the swing. She had counted them one day at recess.

WE LOVE TRIANGLES: EXPLORE SHAPES

The swing seats were squares. She knew because each seat had 4 equal sides and 4 corners shaped like the capital letter L.

For the first time, Rhombi saw that the swing set had a big triangle at each end of the top bar. Rhombi watched her friend Quad swing. She looked at the triangles holding up the swing set. The bases of the triangles at each end never moved.

She also saw that the climbing dome was a bunch of triangles connected by circles.

Triangles were everywhere.

WE LOVE TRIANGLES: EXPLORE SHAPES

After recess, Rhombi's class had center time. Rhombi went to the hamster cage to clean the cedar. She would refill the pet's water, too.

The hamster sat on its nest. The habitat was a cube. It had a red base. The top and sides were clear plastic so kids could see the hamster. The poor rodent looked bored inside its 3D home. Rhombi wished she could give the pet recess. Kids used recess to recharge their brains. The hamster looked like it could use a brain break.

Rhombi smiled and said, "Well, if I can't take you to recess. I can bring recess to you!"

WE LOVE TRIANGLES: EXPLORE SHAPES

Rhombi looked out the window. She saw the playground bridge. She saw the swings. Then Rhombi saw the slides.

“That’s it,” she thought. “I will make a slide for our class hamster.”

Rhombi knew she would need some help. She looked out from the pages of her world and asked, “Friends, will you help me design and make a slide with a sturdy triangle base for our hamster? He weighs about as much as a beanbag.”

WE LOVE TRIANGLES
TALK ABOUT STEAM

What other circles can be found in the paragraph?
once, upon, Rhombi, loved, cool, word, also, out, how, to, for, lot, of

How might Rhombi test a shape to see that it is a square?
Look for a single right angle corner. Measure all sides: they should be equal.

Where does Rhombi see triangles on the playground?
Swing, Slide, Bridge

Why doesn't the swing set tip over when Quad swings?
The base is buried in the ground to make it more stable.

What shape makes the bridge strong enough to hold kids?
Triangles

What does the class do right after recess?
Center time

What is Rhombi's job at today's center?
She is caring for the pet.

What pet does the class have?
hamster

What is the shape of the hamster's habitat?
cube

What color is the cube's base?
red

Why are the other 5 faces clear?
Kids can see the hamster. Hamster can see the kids.

What project does Rhombi want to begin?
She wants to make a slide for the hamster.

TIME TO DESIGN

"Designers, will you create a hamster slide?"

LOOK

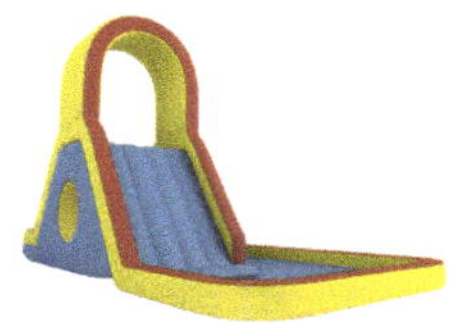

DRAW

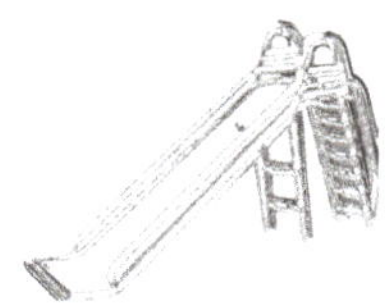

FIND

MAKE

RHOMBI'S ADVENTURES IN 2D
CHALLENGE RUBRIC: WE LOVE TRIANGLES

TASK **Students will use their knowledge of triangles to troubleshoot and build a playground slide that holds 6 ounces of weight.**

	Content	Organization	Presentation
1	• Is well thought out and supports the solution to the challenge or question • Reflects application of critical thinking • Has clear goal that is related to the topic • Is accurate	• Information is clearly focused in an organized and thoughtful manner • Information is constructed in a logical pattern to support the solution	• Format enhances the content
2	• Supports the solution • Has application of critical thinking that is apparent • Has no clear goal • Has some factual errors or inconsistencies (i.e. wheels do not rotate around axle)	• Project has a focus but might stray from it at times (more concerned with form than function) • Information appears to have a pattern, but the pattern is not consistently carried out in the project	• Format does not suit the content
3	• Provides inconsistent information for solution • Has no apparent application of critical thinking • Has no clear goal • Has significant factual errors, misconceptions, or misinterpretations	• Content is unfocused and haphazard • Information does not support the solution to the challenge or question • Information has no apparent pattern	• Presentation appears sloppy and/or unfinished

RHOMBI'S POLYGON ROUND-UP
TRIANGLE

ABOUT THE TRIANGLE

- Triangles are polygons with the least possible number of sides (three).
- The three internal angles of a triangle always add to 180 degrees.
- An equilateral triangle has three sides of equal length and three equal angles.
- The longest side of a right angle triangle is called the hypotenuse, it is always found opposite the right angle.
- Trigonometry is the study of the relationship between the angles of triangles and their sides.
- Triangle shapes are often used in construction because of their great strength.

UNIT 4

TAKE FIVE

TAKE FIVE

Essential Questions

1. How many edges and vertices are present in a regular pentagon?
2. What important monument is built in the shape of a pentagon?
3. How would you draw a pentagon?
4. What are you trying to accomplish in Rhombi's Challenge?

ENDURING UNDERSTANDING

Two dimensional shapes create new shapes.

The length of an object may be discussed using whole numbers.

Organizing information helps in discussion.

Number information may be organized for the purpose of answering questions.

Shapes may be identified by their attributes.

Symmetry means a shape can be folded over edges will match

Asking questions and taking part in discussions increases understanding of a subject.

How many different objects, flora and fauna exist in a pentagonal shaped piece of land? What inventions would help the class? Design and improve signage material. Learn to read infographics. Investigate and create comic strips with 5 panels. Discover the true size of America's famous pentagon.

Students will count to 100 and bundle, compose and decompose numbers. Halves, wholes, and quarters are introduced in folding paper to make pentagons. Discover symmetry and deepen measurement and geometric understanding.

Help Rhombi update the hiking trail signage with sturdy, brightly colored deer crossing signs.

commotion
a hubbub or disturbance

TAKE FIVE
CURRICULUM CONNECTIONS

Rhombi needs to build deer crossing signs.

MATH BACKGROUND - Pentagons

A pentagon is a 5 sided polygon with interior angles that add to 540 degrees. Regular pentagons have edges/sides of equal length and interior angles / vertices of 108 degrees. The edible plant okra is shaped like a pentagon.

ART CONNECTIONS

Draw a pentagon with pencil and ruler. Use the straight edge to draw a line from corner to corner diagonally. Repeat in the opposite corners. Draw a dot in the center.

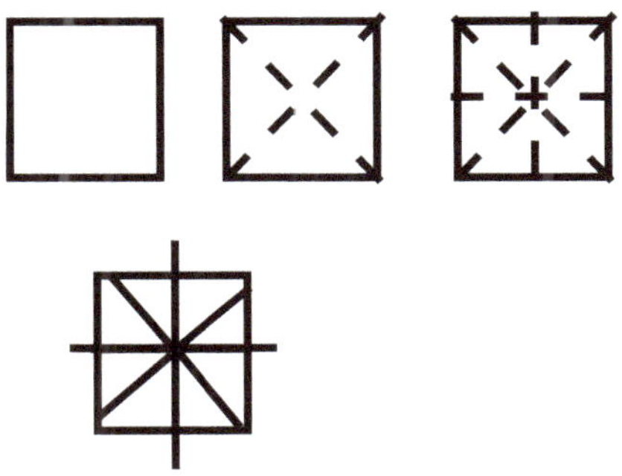

TECHNOLOGY CONNECTION

Julett Morris Elementary Explains Shapes
http://www.schooltube.com/organization/122887/
Students use shapes to build objects.
http://bit.ly/q8OY97
Shapes Song by Peter Weatherall
http://www.youtube.com/user/peterweatherall/videos

HISTORICAL CONNECTIONS

The Pentagon is home to Department of Defense and has three times the floor space of the Empire State Building in New York. According to the website, "Despite 17.5 miles of corridors it takes only seven minutes to walk between any two points in the building." Pentagon Tours explain that there "are 19 escalators, 131 stairways, 284 restrooms, 691 water fountains and enough telephone wire in the building to wrap around the world four and one-half times!"

READING CONNECTION

Introduce each activity and lesson with a book. Books are listed on the Read about STEAM page. Jot additional lists of books from your favorites here.

1.

2.

3.

TAKE FIVE
CURRICULUM CONNECTIONS

COMMON CORE CONNECTIONS

ELA/Literacy
W.K.7 Participate in shared research and writing projects (e.g., explore books by a favorite author and express opinions). (K-ESS2-1)

Mathematics
MP.2 Reason abstractly and quantitatively. (K-ESS2-1)
MP.4 Model with mathematics. (K-ESS2-1)
K.CC.A Know number names and the count sequence. (K-ESS2-1)
K.MD.A.1Describe measurable attributes of objects, such as length or weight. Describe several measurable attributes of a single object. (K-ESS2-1)
K.MD.B.3 Classify objects into given categories; count the number of objects in each category and sort the categories by count. (K-ESS2-1)

NEXT GEN SCI CONNECTIONS

K-2-ETS1-1. Ask questions, make observations, and gather information about a situation people want to change to define a simple problem that can be solved through the development of a new or improved object or tool.
K-2-ETS1-2. Develop a simple sketch, drawing, or physical model to illustrate how the shape of an object helps it function as needed to solve a given problem.
K-2-ETS1-3. Analyze data from tests of two objects designed to solve the same problem to compare the strengths and weaknesses of how each performs.

Rhombi Audio Download and Videos (Available April 2018)

Mod1: Take Five:	Science:	What will you discover in a small piece of land?
Mod1: Take Five:	Technology:	Can students design a new tool?
Mod1: Take Five:	Engineering:	Design and improve signage material.
Mod1: Take Five:	Art & Music:	Investigate and create comic strips with 5 panels.
Mod1: Take Five:	Mathematics:	Discover the true size of America's famous pentagon.
Mod1: Take Five:	ELA Activity:	Read Rhombi's "Take Five." Students illustrate the story.

TAKE FIVE

BUILD STEAM WITH GREAT BOOKS

SCENCE

Grand Canyon
by Cari Meister

One Small Square (caves)
by Donald Silver

Shapes in Sports
by Rebecca Rissman

TECHNOLOGY

Goodnight, Goodnight Construction Site
Sherri Duskey Rinker

Tool, Time, Twist: A Brief History of Tools Through Time
by David Shapiro
Http://www.trafficsign.us

ENGINEERING

Monkey with a Tool Belt
by Chris Monroe

The Kids' Building Workshop
by Craig Robertson

Shapes in Transportation
by Jennifer Rozines and Roy Roy

ART & MUSIC

13 Art Inventions Children Should Know
by Florian Heine

Shapes (Museum of Modern Art)
by Philip Yenawine

Math-terpieces
by Greg Tang

MATH

Pentagon
by Sheila Rivera

How Long or How Wide?: A Measuring Guide
by Brian P. Cleary

ELA

How Margaret E. Knight Became an Inventor
by Marvelous Mattie Emily Arnold McCully

SCIENCE

TAKE FIVE
STEMVESTIGATIONS: STAKE OUT THE PENTAGON

READ **We're Hunting for a Pentagon**

DISCUSS Fallen trees become home to many insects, plants, fungi, and small animals. The forest floor supports thousands of species.

EXPLAIN Forests are unique to each part of the world, but they all have a few things in common. View the virtual forest tour for a glimpse into biomes, habitat, forest creatures, and forestry jobs.

MOVE Play "Going on a Quad Hunt." In the spirit of "Going on a Bear Hunt," sit in a circle with legs folded. Slap legs in a rhythm (slap, slap, clap, clap, slap, slap, clap, clap). Each person in the circle adds a scene to the chant. For a sample, visit Schooltube for videos.
http://bit.ly/rtpbNU or
www.sophie-world.com/activities/singalong-were-going-bear-hunt

INVESTIGATE

What will you find in one small shape of your school yard? Mark off pentagonal areas with sports tape or ribbon around stakes in the ground. No stakes handy? Use pencils in the ground. Use the pentagon pattern in Art Connections to create staked out shapes for examination. For 10 minutes, carefully observe and collect a list of objects (and their shapes) found in each pentagon of earth.

ELA Present findings to the class.

TAKE FIVE
STEMVESTIGATION: STAKE OUT THE PENTAGON

What lives on a pentagon of Earth?
Draw what you find.

We're Hunting for Pentagons

Use tools to collect information.

We're Hunting for Pentagons

Use tools to collect information.

We're going on a 5 hunt. (We're going on a 5 hunt.)
We're looking for a pentagon. (We're looking for a pentagon.)
We're gonna catch a big one. (We're gonna catch a big one.)
I'm not afraid. (I'm not afraid.)
Are you? (No way!)
Here comes the crossing sign.
Better wait for the crossing guard.
Walk briskly and pump your arms.

REFRAIN

We're passing a tall house. It sure is wide and has 5 edges. Let's walk around it.

REFRAIN

We're walking through the soccer field. Uh-oh. Here comes the soccer ball.
We better duck under it…

REFRAIN

We're running through the garden now. It sure is grassy. I see the morning glories and okra. Jump over the rows. Oooohhh. We're all muddy. Shake it off….

REFRAIN

I see a big building. It has 5 sides. It's The Pentagon.
Oh no, what time is it?
Recess starts in 5 minutes. We're going to miss it.
Better get home fast!
Oh no, mud! Shake it off.
Jump the okra and morning glories.
Yikes, duck the soccer ball.
Walk quickly around the tall house.

STOP! The crossing guard says to wait. Okay, we can cross now.
Walk briskly and pump those arms. Hurry! Hurry! Ahhhhh, we made it!

TECHOLOGY

TAKE FIVE
STEMVESTIGATIONS: TOOLS

READ **Tool Time**

DISCUSS Tools are nonconsumable physical objects used to achieve a goal. Tool use has been documented since the beginning of human history, and animals also employ tools. Astronomers use telescopes to see faraway places in greater detail.

EXPLAIN Humans have created tools to simplify work since the stone age. Explore the history of tools from the stone hand axe to the modern toolbox. 20 tools and 2.5 million years of tool making offer a lot to think about.

MOVE Visit the web to locate great songs about construction and tools. Several fun songs that can be acted out while sung.
http://www.prekfun.com
http://www.learninggamesforkids.com/science_songs.html

INVESTIGATE

What activities could be enhanced by the use of an appropriate tool? Students should brainstorm a list of things they do each day that could be easier. What tasks are time-consuming, laborious, physically difficult, or just plain annoying?

Students should break into teams of four to design and develop a tool to achieve a goal. Identify the need. Explain the solution. Compare the idea to other existing tools. List materials to be used in the tool's creation. Gather supplies to create the tools and demonstrate their uses.

ELA Create "advertisements" to sell tool ideas to investors.

TAKE FIVE

STEMVESTIGATION: TOOLS

Tools make life easier. Make a tool for the classroom.

What activities happen in the classroom?

What tools do you use in the classroom?

What problems could you solve in the classroom?

What tool could you create that makes life easier?

On the back of the paper, draw your new tool!
Work with a team to design and build your tool.

Tool Time

Explore the use of tools to solve problems.

Tool Time

Explore the use of tools to solve problems.

Hilda and Poly had a job to do. Grandma wanted the garage organized. She liked everything in its place. The two friends had taken out the trash. They had bagged glass, cardboard, newspaper, glass and cans for recycling. They had even swept and dusted.

The tool bench was still a big mess. Tools were scattered on the table. Tools were dumped in boxes under the bench. Tools were hung in all the wrong pegs on the wall.

Hilda and Poly started by sorting the tools used to move items. They hung hammers and mallets on the peg board over jars of nails in lots of sizes. Star-shaped Philips head and flathead screwdrivers used to tighten or loosen things went into bins by size. #1, #2, and #3 were matched to their same sized screws.

Rhombi put grabbing, twisting, and pulling tools on the pegboard. Pliers were hung from smallest to largest. Wrenches were hung over jars of nuts and bolts.

Rhombi and Quad did not touch the cutting tools. Saws, hacksaws, pincers and wire cutters were left for Grandma to handle later.

Carpenter's levels, rulers, angles, squares, and other measuring tools were hung neatly to the right of the pegboard wall behind the bench. A caliper used to measure very tiny items was placed in its box. Some tools were delicate.

When Hilda and Poly finished the job, they proudly called Grandma to see her shiny garage. The excited old woman clapped her hands and hugged both girls. When Grandma offered them money for doing the job, Hilda refused to take the money.

"No, Grandma," said Hilda. "I'm happy to help!"

ENGINEERING

TAKE FIVE
STEMVESTIGATIONS: MATERIALS

READAD **Creative Design**

DISCUSS "Pentagon" introduces shapes in a repetitive format that makes vocabulary memorable. Pentagons are most widely recognized by students in the national military building and in school crossing street signage.

EXPLAIN The pentagon is a shape with 5 sides, 5 edges, 5 corners, and 5 vertices. A ***regular*** pentagon's edges are all the same length, and its vertices are all the same angle.

MOVE Take a walk out to the "crosswalk" sign.

INVESTIGATE

Crosswalk signs are rigid pieces of metal or plastic that can stand up to strong winds, rain, and changing temperatures. How strong are other materials used in the classroom? Choose 3-4 different materials (fabrics, papers, cardboards, plastic cups, metal pans, etc.) Cut all materials to the same size pentagons.

Each team of students chooses a varying location (shade, under tree, on sidewalk, wall, direct sunlight, etc.). Note temperatures and rainfall for these days. Collect data for a week to a month on a chart with places for the following attributes: color (same, fade1, fade2, fade3, lost all color); thickness (same, worn spots, holes develop, etc.) and texture.

ELA Design other school safety messages for the PENTAGONAL crosswalk sign!

STEMVESTIGATION: MATERIALS

How well do materials stand up to sunlight and rain?

Material	Color	Place	Day 1	Day ___	Day __
Felt					
Plastic					
Fabric					
Wood					

Creative Design

Test materials in design.

How Many Pentagons Can You Count In This Set?

Creative Design

Test materials in design.

Pentagons have 5 edges, 5 edges, 5 edges.
Pentagons have 5 edges,
regular or not.

Regular pentagons have 5 edges
all the same, all the same, all the same.
Regular pentagons have 5 edges
all the same length.

Irregular pentagons have 5 edges
not the same, not the same, not the same.
Irregular pentagons have 5 edges
NOT the same size.

ALL pentagons have 5 angles, 5 angles, 5 angles.
All pentagons have 5 angles
regular or not.

The angles of all pentagons
add up, add up, add up.
The angles of all pentagons
add up to 540 degrees.

5 Edges
5 Angles
500 + 40 Degrees
PEN-TA-GONS!

ART & MUSIC

TAKE FIVE
STEMVESTIGATIONS: STORY IN 5 BOXES

READ

Design Time

DISCUSS

Inventions take many forms. The Utah State Intellectual Properties Services defines an invention as "the discovery or creation of a new material (either a new manufactured product or a new composition or matter), a new process, a new use for an existing material, or any improvements of any of these." Inventions are easy to define but can be difficult to recognize.

EXPLAIN

The art of cartooning was an invention credited to Wilhelm Busch in the late 1800's. Cartoons were listed as "sequential art" with a story told in pictures in strips.

MOVE

Pull several comic strips from the newspaper. Students act out the panels while other try to identify the matching strip.

INVESTIGATE

Katzenjammer Kids was first printed in the New York Journal in 1897. What are students' favorite cartoons and comic strips? Cut out cartoons from the newspaper to share. Students read the cartoons (or have them read) to get ideas. What do all cartoons seem to have in common? What aspects of each cartoon are unique? Create cartoons to support a science or math lesson. Characters might -for example- have to solve a math puzzle or mystery. They might have to explain a scientific concept being learned in school.

ELA

Create a 5-panel comic strip cartoon book to share with other classes.

TAKE FIVE
STEMVESTIGATION: STORY IN 5 BOXES

Create a comic strip story in five parts.

TITLE

Design Time

Explore polygons in art.

Design Time

Explore polygons in art.

Young man, you amaze me. *--Albert Einstein, physicist (1938)*

Buckminster Fuller said, "Whenever I draw a circle, I immediately want to step out of it." Buckminster Fuller was an architect and an engineer born in 1895. He was also a designer and an inventor. Fuller wrote 28 books on subjects in math and science. He had 47 special degrees from different universities.

Fuller was someone that cared about making the world better. He was also a humanitarian, someone that wants to improve life for other people. Each year the Buckminster Fuller Institute offers a $100,000 prize to "scientists, students, designers, architects, activists, entrepreneurs, artists and planners from all over the world using innovative solutions to solve some of humanity's most pressing problems." From a floating doctor's office to a new plan for food production, Fuller's prize helps people with great ideas afford to put their plans in action.

This inventor was interested in using the planet's resources more efficiently. He invented a geodesic dome made of polygon shapes. The dome uses fewer materials and can be made bigger or smaller to meet the needs of many different types of uses. Gardens, homes, and museums are shaped like geodesic domes.

When Buckminster Fuller first imagined the special dome, he thought of it as humans working like a crew to survive on our planet, "spaceship earth." Epcot Center in Florida created a domed building inspired by Fuller's idea.

Domes like these are made of triangles. As more triangles are added, the shape gets stronger. No other 3-dimensional shape can be so big without extra supports inside the building. The larger it is, the stronger the dome becomes. These domes are so simple that an entire house can be built in a single day. They are so strong that geodesic domes can even survive hurricane force winds!

Can you see the triangles in the playground dome?
Can you see diamonds in the playground dome?
Can you see hexagons in the playground dome?

MATHEMATICS

TAKE FIVE
STEMVESTIGATIONS: THE PENTAGON

READ **The Pentagon, USA**

DISCUSS The Pentagon is one of America's best known buildings. Its design was chosen to fit an unusually shaped piece of land. When the site changed, architects and engineers kept the pentagonal shape of the building for its new location.

EXPLAIN According to GreatBuildings.com, each outer edge has a length of 921feet. Other numbers include:

- Floor area of 6.5 million square feet or 34 acres
- 3.7 million square feet are used for offices
- 80 million dollar construction cost in 1943 dollars
- 17.5 miles of corridors

MOVE Each edge of the Pentagon measures 921 feet. Walk 921feet. Walk it 5 times to simulate walking the perimeter of this huge building.

INVESTIGATE

As a class, create a scale model of the Pentagon building with blocks and other classroom construction materials. If each outer wall measures 921 feet, adjust your classroom or playground structure to measure a scaled down 9.21 feet on each edge. How will the class make sure the structure is actually a regular pentagon with equal angles? (Hint: They can print and cut out a paper sized pentagon. Lay the paper at your large pentagon's center. Assemble large walls parallel to the small pentagon's edges.) What materials will you use?

ELA Rethink your comic strip. Add facts to the original comic.

TAKE FIVE
STEMVESTIGATION: THE PENTAGON

Walk The Pentagon. Test its angles with a pattern.

The Pentagon, USA

HOLDS A LOT OF CARS

= 10,000

Aerial Image by David B. Gleason from Chicago, IL (The Pentagon) via Wikimedia Commons

COUNT on the PENTAGON

1 formal dining room
2 cafeterias
3 days to design the building
4 thousand 2 hundred clocks
5 edges and vertices in a pentagon
6 snack bars
7 minute walk corner to corner
8 decades since building opened
9 hundred 21 feet per outer edge
10 thousand visitors a year

5 Stories Tall
+2 below ground

5 Armed Services
housed on site

5 Acre Central Plaza
"Ground Zero"

Executive Order 8802 let The Pentagon open all its 284 bathrooms to all employees, black or white. For a time, The Pentagon was the only non-segregated building in Virginia.

NUMBER OF MONTHS FROM GROUND BREAKING TO OPENING CEREMONY

	JAN	FEB	MAR	APR	MAY	JUNE	JULY	AUG	SEPT	NOV	DEC
1941									■	■	■
1942	■	■	■	■	■	■	■	■	■	■	■
1943	■										

Reference Material
January 15, 2013 9 Things You May Not Know About the Pentagon By Barbara Maranzani via History.com
Vogel, Steve. The Pentagon: A History. New York: Random House, 2007. ISBN 1-400-06303-5 (pp. 199-200, 216-217)
History of the Pentagon — U.S. DOD info, archived at Archive.org.
Pentagon Web Site — By the U.S. Department of Defense.
http://en.wikipedia.org/wiki/The_Pentagon and Wikimedia Commons

Graphic Information

Explore infographics in mathematics.

Infographic is short for information graphic. An infographic uses pictures and words to make lots of information and numbers easy to grasp. Look at this infographic showing information about The Pentagon. You could read the paragraphs that follow, or you can use an infographic to show the same information.

The Pentagon was designed in just 3 days and built far faster than most government buildings. It was also built to hold a lot of cars. Parking was planned for 10,000 cars. 16 months passed from the first day of construction until employees parked all those cars in the lots. The construction trucks, earth movers, and builders showed up for work on September 11, 1941. The Pentagon was completed in January of 1943. The building opened almost 8 decades ago.

The building serves a lot of needs. Very special visitors are invited to parties in the formal dining room. Most armed services workers eat in one of the 2 cafeterias. 5 different armed services use the building. 10,000 or more visitors use any of the 6 snack bars.

Finding a bathroom in The Pentagon is easy. When the building was planned in 1941, Virginia state law still said that black people and white people had to use different bathrooms. The pentagon was built with 284 bathrooms. By order of the President, none of those bathrooms were ever separated. Employees and visitors of all backgrounds used any of the Pentagon's public bathrooms. For a time, The Pentagon was the only non-segregated building in the state of Virginia.

Getting from place to place in the huge building takes time. Each outer edge measures 921 feet. There are 5 stories full of offices and a 5 acre central area called "ground zero." The total square footage of the The Pentagon is more than 5 million square feet. Your school probably measures less than 50,000 square feet. Walking from one corner to another corner of The Pentagon takes nearly 7 minutes. How many seconds pass if you walk from one corner of your building to another corner of your building?

Which format seems easiest to use? Which format gives information quickest? Which format offers more of a story? Do you like the paragraphs or graphic style best?

DOT TO DOT: GEO PAPER GAME

MATERIALS

You will need an overhead projector or smart board setup. Copy the dot paper on a transparency. You will need appropriate marking tools (transparency markers or smart mouse setup).

SETUP

Project dot paper on a screen or whiteboard.

INSTRUCT THE PLAYERS

Taking turns, the first player connects two dots with either a horizontal or vertical line. The other players take their turns doing the same. Their lines may touch other player's lines, but they don't have to. When a player draws the fourth line to create a box, the team claims the box and writes the first letter of its team name inside the box.

STUDENTS BREAK INTO TWO CLASS TEAMS OR SMALL TEAMS

Play. Once all of the possible horizontal and vertical lines in the grid have been turned into boxes, each team counts the number of boxes it owns. The team with the most boxes wins.

UNIT 4 CHALLENGE

TAKE FIVE

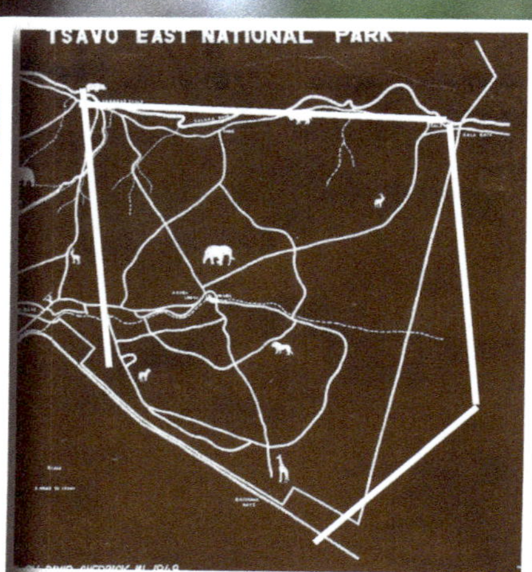
TSAVO EAST NATIONAL PARK

TAKE FIVE

Rhombi, Quad and Poly lace up hiking boots and fill water bottles. They each have trail maps. They wear long pants and long sleeved shirts. Mosquitoes and other insects can be trouble on a hiking trail. Quad hands out sunscreen, and they put on hats.

The kids will hike trails to look for deer, rabbits, snakes, raccoons, squirrels, birds, and turtles. The kids open a map. The park has 5 edges and 5 vertices.

"Hey, the trails make a pentagon like the shapes on my soccer ball," Poly shouts!

Quad laughs but shushes Poly. "We will never see any animals if you keep shouting like that."

They choose a trail leading up to the deer meadow. With any luck -and some quieter voices- they might actually see some animals in the woods. Starting up the trail, Rhombi points to morning glories. The flowers are covered in shiny morning dew. The pentagonal flowers are beautiful.

TAKE FIVE

TAKE FIVE

"Look at the deer," whispers Poly.

Peeking from the edge of the woods are several deer. Their tracks show that many deer have used this place on the path. The kids watch from their place down the hill. They are perfectly still and silent. As Rhombi, Quad, and Poly watch, five deer move across the path.

Suddenly, there is a commotion. Mountain bikers speed down the trail toward the deer. Poly shouts at the deer to move faster. She shouts at the bikers to stop!

The bikes have a right to be on the trail. They have permits and are being safe, but an accident could happen any second! The mountain bikes skid to a stop. Gravel and dirt fly into the air. The last deer bounds away into the trees.

"Whew! There should be a sign or something if this is a place the deer walk a lot," says one biker. She is breathing hard. "You know, like a school crossing sign or something."

TAKE FIVE

TAKE FIVE

The two groups wave goodbye. Poly looks at Rhombi and Quad. She says, "You know. They are right. The trail SHOULD have some sign for deer trails. That way we'll know where to look, and the bikers will know to slow down in that area."

At the end of the trail, all three friends sit down to think about sign making. They have seen the school crossing signs. Each sign has five edges and five vertices. It's a regular pentagon, so all the edges are the same length. It's bright yellow and easy to see.

Rhombi decides to ask her innovative friends for help on this project.

She looks out from the pages of her world and asks, "Designers, will you create deer crossing signs like the ones outside your school? They have to stand up to wind and rain and be bright enough to see."

TAKE FIVE
TALK ABOUT STEAM

What shape helped the cart roll?
circle

What innovative ways can you check for squares?
Use the corner of a notebook paper. Make a "t" square.

What shape made the hamster slide strong?
triangle

What steps do the kids take to be safe in the woods?
They wear long sleeves, long pants, hats, sunscreen and water bottles. They take maps.

What flowers have a pentagonal shape?
morning glories

What animals did the kids see?
deer

Why did the kids decide to post a "deer crossing" sign?
deer and some mountain bikers nearly ran into each other.

What is the shape of the 5-sided sign?
pentagon

Name at least 3 objects that have regular pentagons? star fruit cross section, crosswalk sign, Pentagon building in Washington, D.C., USA

What are the rules for a regular pentagon?
5 edges of equal length with 5 vertices of equal angle

TIME TO DESIGN

"Builders, will you design and make sturdy crossing signs?"

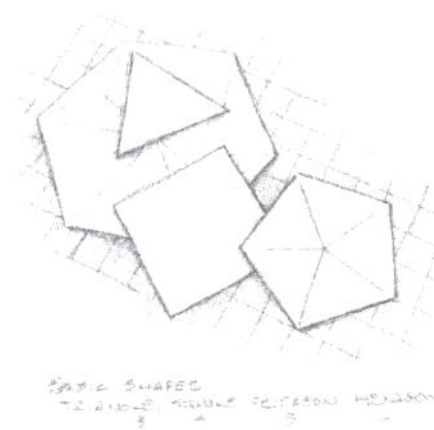

LOOK

DRAW

FIND

MAKE

RHOMBI'S ADVENTURES IN 2D
UNIT 4 RUBRIC: TAKE FIVE

TASK **Students will use their knowledge of pentagons, materials, and weather to design, build, and troubleshoot a deer crossing sign.**

	Content	Organization	Presentation
1	• Is well thought out and supports the solution to the challenge or question • Reflects application of critical thinking • Has clear goal that is related to the topic • Is accurate	• Information is clearly focused in an organized and thoughtful manner • Information is constructed in a logical pattern to support the solution •	• Format enhances the content
2	• Supports the solution • Has application of critical thinking that is apparent • Has no clear goal • Has some factual errors or inconsistencies (i.e. wheels do not rotate around axle)	• Project has a focus but might stray from it at times (more concerned with form than function) • Information appears to have a pattern, but the pattern is not consistently carried out in the project •	• Format does not suit the content
3	• Provides inconsistent information for solution • Has no apparent application of critical thinking • Has no clear goal • Has significant factual errors, misconceptions, or misinterpretations	• Content is unfocused and haphazard • Information does not support the solution to the challenge or question • Information has no apparent pattern •	• Presentation appears sloppy and/or unfinished

RHOMBI'S POLYGON ROUND-UP
PENTAGON

ABOUT THE PENTAGON

- A pentagon is a 5 sided polygon with interior angles that add to 540 degrees.
- Regular pentagons have sides of equal length and interior angles of 108 degrees.
- The US Department of Defense headquarters is named 'the Pentagon'.
- The edible plant okra is shaped like a pentagon.

TINY HOUSE ON WHEELS

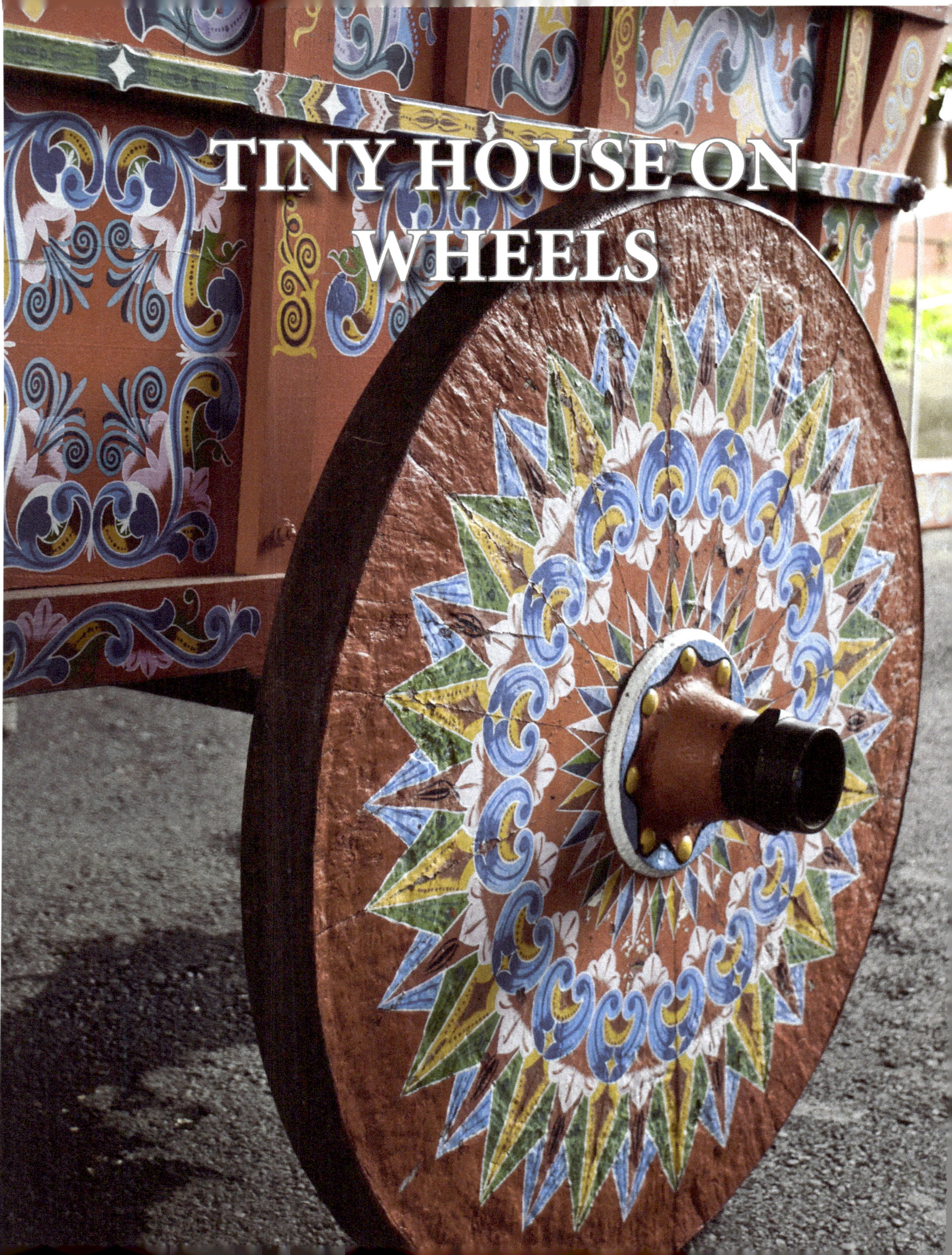

TINY HOUSE ON WHEELS

Essential Questions

1. What tools should be used to measure ounces & feet?
2. What units should be used to measure mass/weight and distance/length?
3. Can you talk about shapes using attributes?
4. What will happen to the shapes in your design when they are transformed?

ENDURING UNDERSTANDING

Numbers can represent quantity, position, location, & relationships.

Computation involves taking apart and bundling numbers

Objects may be described by their attributes.

Points, lines, and planes are the basis for geometry.

Transformation of an object does not change its shape.

Objects have distinct attributes that can be measured.

Standard units provide common anchors for discussion.

While this project may be approached as a stand-alone investigation, it also acts as an authentic assessment. Students will apply knowledge of physical science, earth science, materials, attributes, geometry, numeracy and measurement to design and build a scale model.

Rhombi and her friends need a covered wagon or recreational vehicle to use on their camping trip. The wagon or cart should hold all the kids' gear and function as a sleeping area in case of rain. As a scale model, the student project should hold 6 ounces of "gear" and roll for at least 3 feet before losing integrity.

Logbooks should contain drawings, designs, and "notes" about the project.

Students should not worry about the motor as this is a model. Explain that the full-size vehicle would have an engine or be pulled behind a vehicle.

Rhombi and her friends are planning a camping trip for Friday night. Quad, Poly and Rhombi have made lists of things they will need. Rhombi looks over the list one more time.

	RHOMBI	QUAD	POLY
sleeping bags	x	x	x
apples and bananas	x		x
water bottles	x	x	x
tortilla chips and salsa	x	x	x
extra sweatshirts	x	x	x
flashlight	x	x	x
tent or wagon	?	?	?
a way to move stuff	?	?	?

They have sleeping bags, water, and snacks. They each have one flash light and extra batteries. Each kid has a warm sweatshirt for the cold part of the night. The kids have almost all their gear ready.

Quad, Rhombi, and Poly may have their gear packed, but they are missing two important things.

TINY HOUSE ON WHEELS

Rhombi looks at the list and sits down to think. She has been reading a book about the early settlers. They moved across the country in covered wagons. Each wagon held what a family would need to start a whole new life. The wagon was also the family's home on the long trip West.

Rhombi had an idea. She was sure her friends were innovative enough to design a sort of covered wagon. She, Poly, and Quad could use the wagon to carry their supplies. They could also sleep in the wagon if the weather turned rainy.

Rhombi looked out from the pages of her world and asked, "Innovators, will you design and build a model of a wagon we could use for camping? We will need to move three people and all of our supplies. Your scale model will need to carry at least 6 ounces of weight for a length of 3 feet."

RHOMBI'S ADVENTURES IN 2D
TALK ABOUT STEAM

What shape helped the cart roll?
circle

What innovative ways can you check for squares?
Use the corner of a notebook paper. Make a "t" square.

What shape made the hamster slide strong?
Triangle

What are Rhombi, Quad and Poly planning to do?
They will go camping.

How does Rhombi plan for the trip?
She makes a checklist.

What are the friends missing?
They are missing a tent and a way to move their gear.

What would Rhombi like her designer friends to create?
She needs a wagon to use as gear cart and sleeping solution.

How have our previous projects helped prepare for this one?
Round wheels and axle combinations made a cart.
We can check the square corners of the wagon with a square tool.
Using triangles for support will make a project stronger.

What steps should designers take in planning for the project?

- Look for inspiration in objects that already exist.
- Sketch and draw several ideas.
- Find materials that might work.
- Build the design.
- Troubleshoot the project.
- Make any changes that might be needed.

TIME TO DESIGN

"Builders, will you design and make a sleeping wagon?"

Draw or list information you already know on this topic.

Think about old projects. Look at pictures of some homes and vehicles.

TIME TO DESIGN

"Designers, will you create make a sleeping wagon?"

Draw or list some ideas.

TIME TO DESIGN

"Designers, will you create a sleeping wagon?"

Draw or list the things you will use.

TIME TO DESIGN

"Designers, will you create a sleeping wagon?"

What tools will you use?
Who will help?

RHOMBI'S ADVENTURES IN 2D
EXTENSION RUBRIC: TINY HOUSE ON WHEELS

TASK **Students will use their knowledge of polygons to troubleshoot Rhombi's planned camping trip and build a rolling covered wagon.**

	Content	Organization	Presentation
1	• Is well thought out and supports the solution to the challenge or question • Reflects application of critical thinking • Has clear goal that is related to the topic • Is accurate	• Information is clearly focused in an organized and thoughtful manner • Information is constructed in a logical pattern to support the solution	• Format enhances the content
2	• Supports the solution • Has application of critical thinking that is apparent • Has no clear goal • Has some factual errors or inconsistencies (i.e. wheels do not rotate around axle) •	• Project has a focus but might stray from it at times (more concerned with form than function) • Information appears to have a pattern, but the pattern is not consistently carried out in the project •	• Format does not suit the content
3	• Provides inconsistent information for solution • Has no apparent application of critical thinking • Has no clear goal • Has significant factual errors, misconceptions, or misinterpretations	• Content is unfocused and haphazard • Information does not support the solution to the challenge or question • Information has no apparent pattern •	• Presentation appears sloppy and/or unfinished

APPENDIX

APPENDIX
CUBE NET

INSTRUCTIONS: Print this net for use in a center or during STEMVESTIGATIONS.

APPENDIX
PYRAMID NET

INSTRUCTIONS: Print this net for use in a center or during STEMVESTIGATIONS.

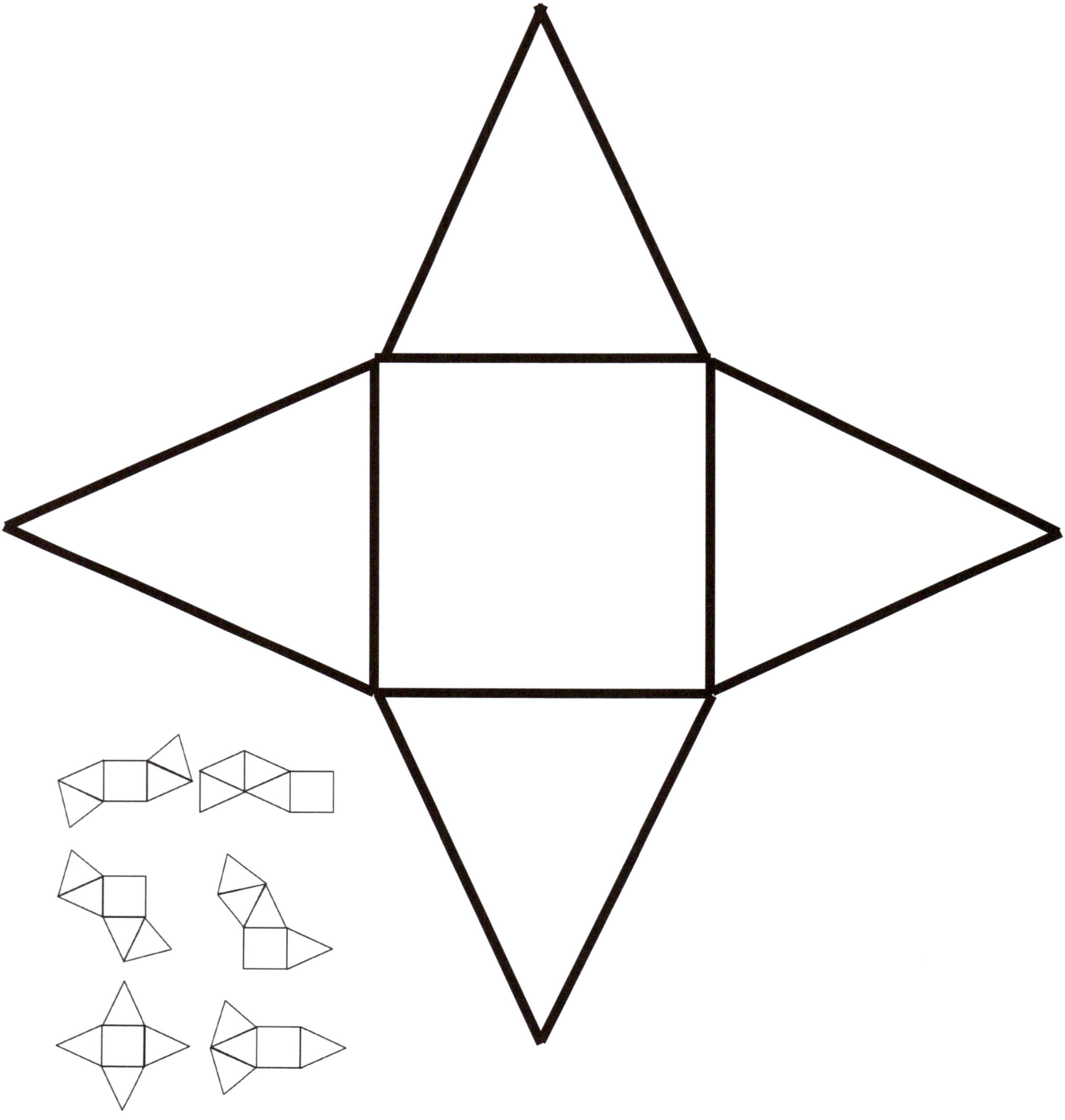

RHOMBI'S POLYGON ROUND-UP
CIRCLE

ABOUT THE CIRCLE

- In the math world, circle refers to the boundary of the shape. "Disk" is used to refer to the whole shape, including the inside.
- A straight line from the center of a circle to the edge is called the radius.
- A straight line that passes from one side of a circle to the other through the center is called the diameter.
- The distance around the outside of a circle is called the circumference. All points on the edge of a circle are the same distance to the center.
- The value of Pi (π) to 2 decimal places is 3.14, it comes in handy when working out the circumference and area of a circle.

RHOMBI'S POLYGON ROUND-UP
TRIANGLE

ABOUT THE TRIANGLE

- Triangles are polygons with the least possible number of sides (three).
- The three internal angles of a triangle always add to 180 degrees.
- An equilateral triangle has three sides of equal length and three equal angles.
- The longest side of a right angle triangle is called the hypotenuse, it is always found opposite the right angle.
- Trigonometry is the study of the relationship between the angles of triangles and their sides.
- Triangle shapes are often used in construction because of their great strength.

RHOMBI'S POLYGON ROUND-UP
RECTANGLE

ABOUT THE SQUARE

- A square is a polygon with 4 sides of equal length and 4 right angle corners (90 degree corners).
- Because it has 4 sides of equal length, a square is a regular quadrilateral.
- A square is also a rectangle with equal sides and a rhombus with right angles.
- The perimeter of a square is 4 times the length of one side.
- Opposite sides of a square are parallel.
- The internal angles of a square add to 360 degrees.
- A square has 4 lines of reflectional symmetry.

RHOMBI'S POLYGON ROUND-UP
PENTAGON

ABOUT THE PENTAGON

- A pentagon is a 5 sided polygon with interior angles that add to 540 degrees.
- Regular pentagons have sides of equal length and interior angles of 108 degrees.
- The US Department of Defense headquarters is named 'the Pentagon'.
- The edible plant okra is shaped like a pentagon.

RHOMBI'S POLYGON ROUND-UP
HEXAGON

ABOUT THE HEXAGON

- A hexagon is a 6 sided polygon with interior angles that add to 720 degrees.
- Regular hexagons have sides of equal length and interior angles of 120 degrees.
- Beehive cells are hexagonal.

RHOMBI'S POLYGON ROUND-UP
HEPTAGON

ABOUT THE HEPTAGON

- A heptagon is a 7 sided polygon with interior angles that add to 900 degrees.
- Regular heptagons have sides of equal length and interior angles of 128.57 degrees.
- The British 50 and 20 pence coins are curved heptagons.

RHOMBI'S POLYGON ROUND-UP
OCTAGON

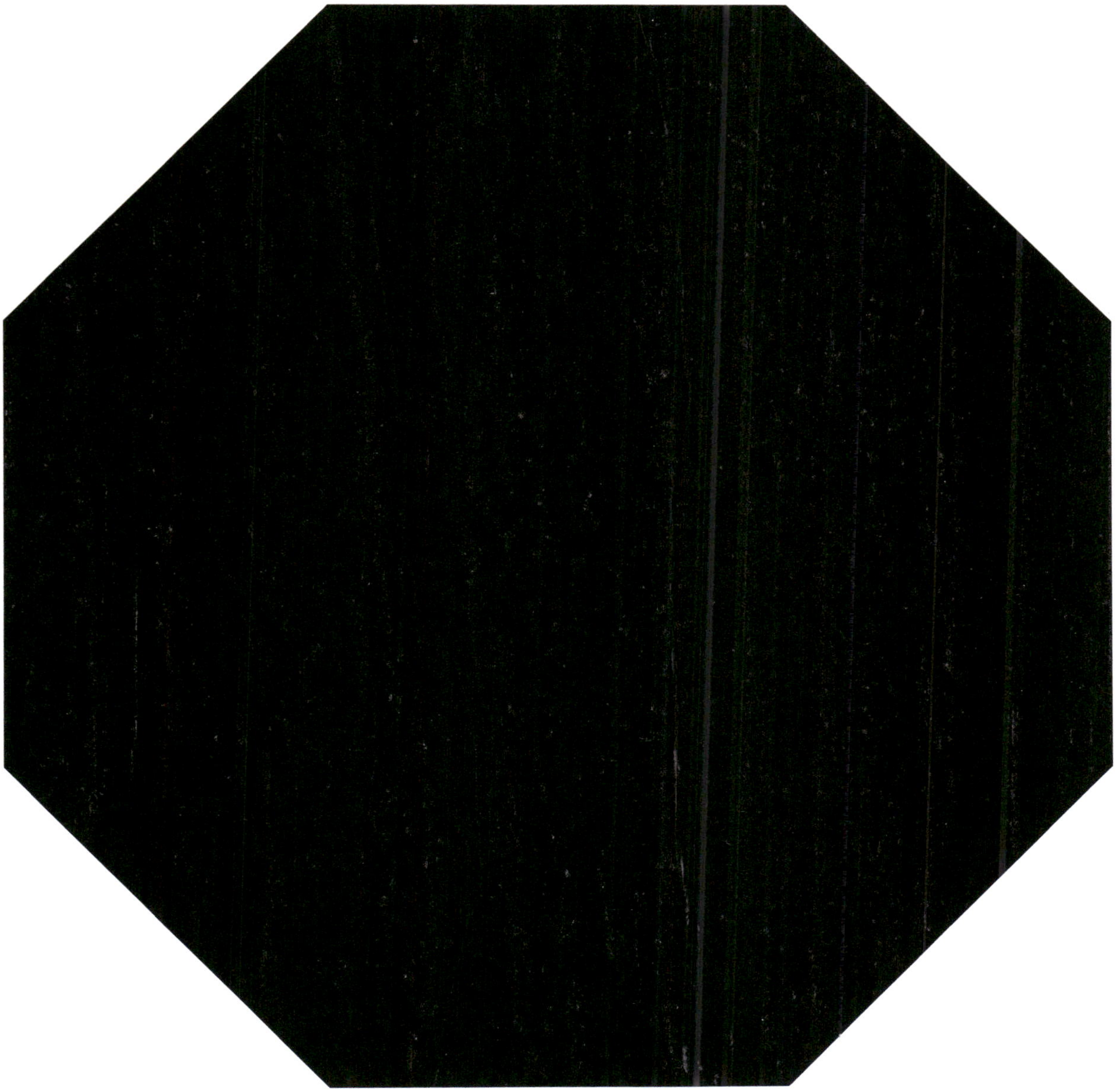

ABOUT THE OCTAGON

- An octagon is an 8 sided polygon with interior angles that add to 1080 degrees.
- Regular octagons have sides of equal length and interior angles of 135 degrees.

RHOMBI'S POLYGON ROUND-UP
NONAGON

ABOUT THE NONAGON

- A nonagon is a 9 sided polygon with interior angles that add to 1260 degrees.
- Regular nonagons have sides of equal length and interior angles of 140 degrees.

RHOMBI'S POLYGON ROUND-UP
DECAGON

ABOUT THE DECAGON

- A decagon is a 10 sided polygon with interior angles that add to 1440 degrees.
- Regular decagons have sides of equal length and interior angles of 144 degrees.

RHOMBI'S POLYGON ROUND-UP
TEACHER'S BACKGROUND

GENERAL INFORMATION ON POLYGONS

A polygon is a geometric figure in two dimensions with three or more sides. The name comes from two Greek words, poly, meaning "many," and gon, meaning "angle." A polygon always has as many angles as it has sides. Polygons are named to indicate the number of sides or angles they contain. Thus, a hexagon has six (hexa- means "six") sides and six angles. The sum of a polygon's exterior angles is 360°. To find the internal angles of a polygon use this formula:

Multiply 180 by (the number of sides minus 2). A hexagon has 6 sides. 6 - 2 = 4

Hexagon is 180° x 4 = 720°

POLYGON TERMINOLOGY

Parts and properties of polygons.

Side: Any one of the straight lines that make up the polygon.

Vertex: A point where any two of the sides of a polygon meet to form an angle.

Angle: A figure formed by the intersection of two sides.

Diagonal: A line that joins any two nonadjacent (not next to each other) vertices.

Perimeter: The sum of the length of all sides.

Area: The space enclosed within the polygon.

TYPES OF POLYGONS

Equilateral: A polygon in which all sides are equal in length.

Equiangular: A polygon in which all angles are the same size.

Regular: A polygon that is both equilateral and equiangular. This is the one most often used in schools.

EXAMPLES OF POLYGONS

The most common kinds of polygons include:

Parallelogram: A quadrilateral (four-sided figure) in which both pairs of sides are parallel and equal.

Rhombus: A parallelogram in which all four sides are equal.

Rectangle: A parallelogram in which all angles are right angles.

Square: A rectangle in which all four sides are equal.

SUM of INTERNAL ANGLES

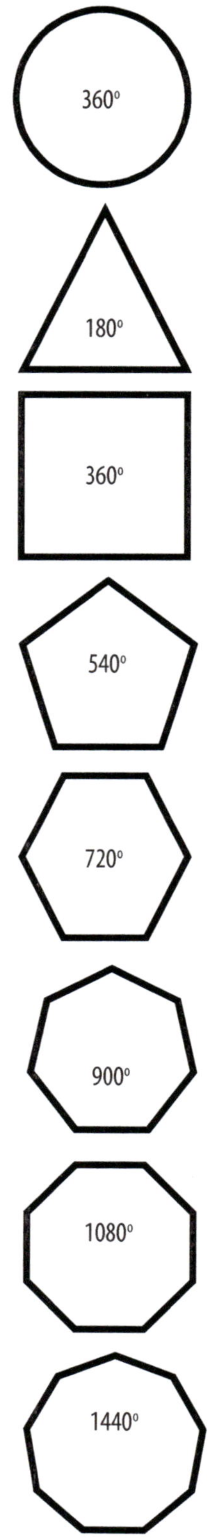

mindbugs

Rhombi's Adventures in 2D

MindBugs in Measurement

MODULE 1: FIRST GRADE

MindBugs in Measurement:

Read Across the Curriculum

Think about Time

Why are there 60 seconds in a minute, 60 minutes in an hour, 24 hours in day or 24 time zones?

Think About History
Ancient Babylonians used base 60. They counted all the way to 59 before bundling 60 items. This base 60 system is still used in maps and clocks. When telling time, you count 1 minute to 59 minutes before bundling 60 minutes into 1 hour.

Time zones as we know them did not exist until the 1840s. 25 countries were involved in the decision to place the Prime Meridian so that it passes through Greenwich, England. In 1884, a vote took place in which 22 countries voted yes; 1 no; and 2 did not vote at all. More countries voted yes than no.

Think About Numbers

Units for measuring time help us organize our lives. As people tried to make smaller and more manageable units to describe and quantify the passing of time, the light and dark that defines each day provided a natural starting point.

The Earth rotates every period of light and dark (day and night). The Earth turns about its axis 360 degrees, or a full circle, in each rotation. The distance around the Earth has been broken into 24 even parts of 15 degrees each. This makes 24 time zones. Each time zone is measured with longitude lines on a map starting at 0 degrees at Greenwich, England.

MindBugs in Measurement:

Read Across the Curriculum

Talk about Time

To understand the history of timekeeping, you'll need to go back over 5,000 years to an era when the cycles of the rising and setting sun, moon, and stars were the natural tools for keeping track of time. Ancient Egyptians used shadows cast from towers called obelisks to mark time. As the sun appeared to move across the sky from east to west, the pole's shadow moved around its base. Morning, noon and afternoon markers were placed around the base. Since the sundial always measured 12 hours of daylight, the ancient Egyptian concept of an hour changed according to season. With longer days in the summer and shorter days in the winter, Egyptian summer hours measured longer than winter hours.

The sand glass, invented some time before the 14th century, used the controlled movement of sand from one chamber to another as a sort of timing device. These devices are more like stopwatches than clocks.

A turning point in keeping time was reached with the invention of the mechanical clock. Mechanical clocks depend on a motion of vibration that is constant and regular to count off the units of time. Early mechanical clocks used the laws of gravity to operate (like Grandfather Clocks). Gravity acted on a heavy weight suspended on a string: a pendulum. The other end of the string was attached to a gear shaft that, as it turned, also turned the main wheel of the clock.

Beginning in the 16th century, spring-driven clocks used coiling and uncoiling of a spring to drive the clock mechanism. Spring-driven clocks replaced the need for a heavy weight hung in a bell tower. Springs allowed smaller clocks to be made for use in the home.

MindBugs in Measurement:

Read Across the Curriculum

Read about Time

SCIENCE FROM A SICKBED

Descartes is given credit for developing the system of coordinate points. A story tells of Descartes lying in bed during an illness, watching a fly move on the ceiling. He realized that he could describe the fly's location at any time using lines "drawn" together from two sides of the ceiling. It was this insight that supposedly led him to develop the grid or coordinate system we still use today.

WHAT'S THE POINT?

A coordinate point is usually written in parentheses and describes the x and y coordinates of a point. (2, 5) on a graph might tell the story of 2 rabbits that each ate 5 carrots. (2,7) on the same graph might tell us that 2 rabbits ate 7 carrots.

To plot a point, move across the x axis to 2, then up the y axis to 5. Where the lines meet, you'll place a dot.

Reading stories on a graph helps scientists, mathematicians, engineers, and statisticians understand the same story even if the storyteller is not present.

FINDING COORDINATE PLANES

Look up. Your classroom ceiling is a perfect coordinate plane. The ceiling tiles form a wonderful grid for plotting x and y. Hang a toy from the ceiling, label your dependent and independent variables, and then ask friends or classmates to tell you how to reach the item using Descartes' coordinate system!

MindBugs in Measurement:

Read Across the Curriculum

Inch by Inch

According to encyclopedia.com, the inchworm is actually the "larvae of moths of the family Geometridae, a large…group with over 1,200 species indigenous to North America." These North American baby moths have become famous as characters in children's stories over the past 200 years.

Inchworms get their unusual name from the length of the body fully stretched. They really ARE about one inch long. According to the Collaborative International dictionary of English, "inchworms" are also called "measuring worms, spooners and loops." These names refer to the looping motion the larvae makes as it travels across the oak tree gobbling up leaves. The hungry inchworms first bring the rear end of the body forward, forming a loop, then move the front part of the body.

An inchworm's smooth, hairless body usually measures about 1 in. or 2.5 cm long. They are green, brown or black and may have lumpy projections that mimic tree twigs. If you scare an inchworm, it may stand still and tall (as tall as an inchworm can get) making them look even more like waving oak twigs.

MindBugs in Measurement:

Read Across the Curriculum

Let Your Fingers Do the Measuring

For thousands of years, the human body has acted as a ruler for measuring. A finger's or hand's width and the length of your foot were all accepted ways to measure length. In the time of England's Round Table, when knights in armor fought battles on horseback, an inch was just the width of a man's thumb. How confusing that must have been! Compare your thumb to someone else's in the room, and imagine trying to build a castle using all the different measurements of a thumb's width.

THE IRON ULNA

Sometime after the Magna Charta was signed in the Thirteenth Century, King Edward I of England took a giant leap in measurement. King Edward knew that having everyone use the same measurement for an inch was important. He decreed that the inch should be one thirty-sixth of a yard. A permanent measuring stick made of iron served as a master standard yardstick for the entire kingdom. This master yardstick was called the "iron ulna", after the bone of the forearm.

AMERICA'S BRONZE BAR

Immigrants to the New World brought with them all kinds of different standards for measuring the inch and the yard. Finally, after some confusing decades, Americans decided that the inch and yard would be measured against the European metric system. America created a bronze bar that's yard measured 82 inches. The bar has been replaced at least once since it's first use in 1832.*

**http://www.cftech.com/BrainBank/OTHERREFERENCE/WEIGHTSandMEASURES*

MindBugs in Measurement:

Read Across the Curriculum

Using the Stopwatch

IN THE BLINK OF AN EYE

Can you use a stopwatch to time an eye blink? Try it. From measurements taken over time, we know that an eye blink takes place in a little over 15 hundredths of a second. That means you've cut a single second into 100 parts. The eye blink is just 15 of those 100 parts. Basically, a blink happens This means that if your reaction time is greater than 0.15 seconds, the blink is over before you can push the buttons.

Because people blink, the average person spends about 23 minutes in darkness ever day. This time in darkness may be why the brain has learned to "fill in the blank spaces" for us on occasion. It may be why "eye witness" testimony is not considered the best source of evidence. It may explain why several witnesses to the same event will have "seen" different things.

SLOW DOWN IN THE CAR

According to the book called Newton Rules Biology, messages move through your nervous system at about 160 feet per second. This speed translates to approximately 100 miles per hour, which explains why driving at high speeds is so dangerous. The speed on a highway is about 65 miles per hour. The speed in a school zone is usually 35 miles per hour.

If a driver is not anticipating the next move or looking ahead for other drivers in trouble, reaction speed can be slower than the speed of the car. Making the fast moves to avoid trouble may not be physically possible.

RHOMBI'S ADVENTURES IN GEOMETRY
RECOGNIZE SHAPES

Color the square.	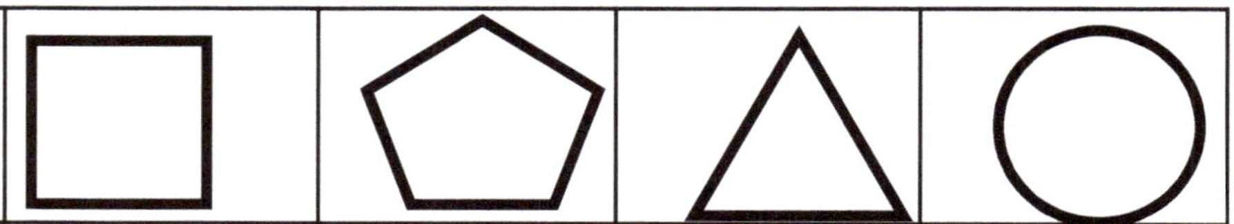
Color the triangle.	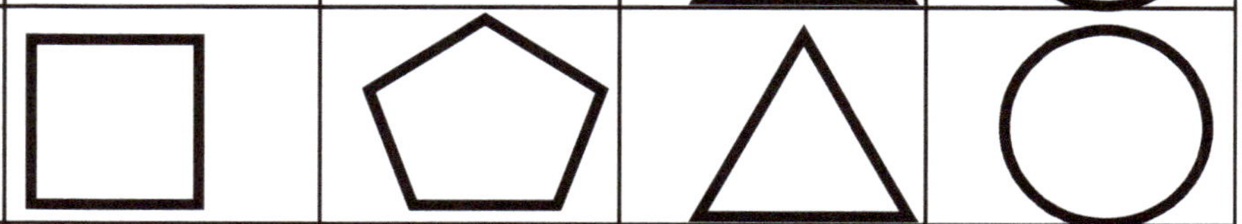
Color the rectangle.	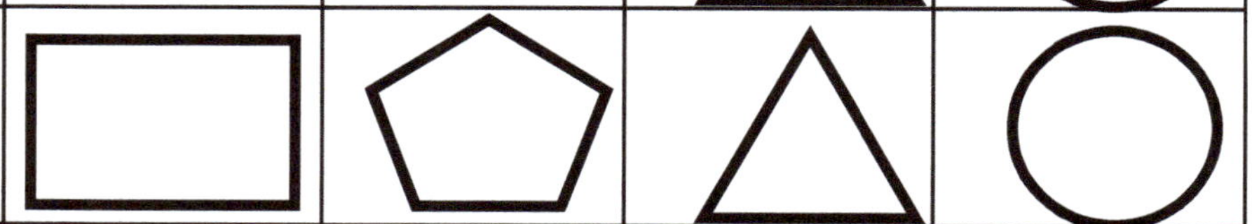
Color the circle.	
Color the pentagon.	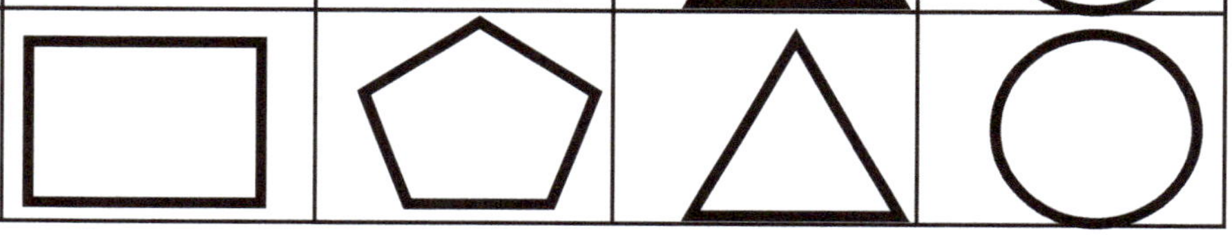

RHOMBI'S ADVENTURES IN GEOMETRY
REPRODUCE SHAPES

Draw a square.	
Draw a triangle.	
Draw a rectangle.	
Draw a circle.	
Draw a pentagon.	

RHOMBI'S ADVENTURES IN GEOMETRY
MOVE FROM 2D TO 3D

Mark the shape you see on a cube.

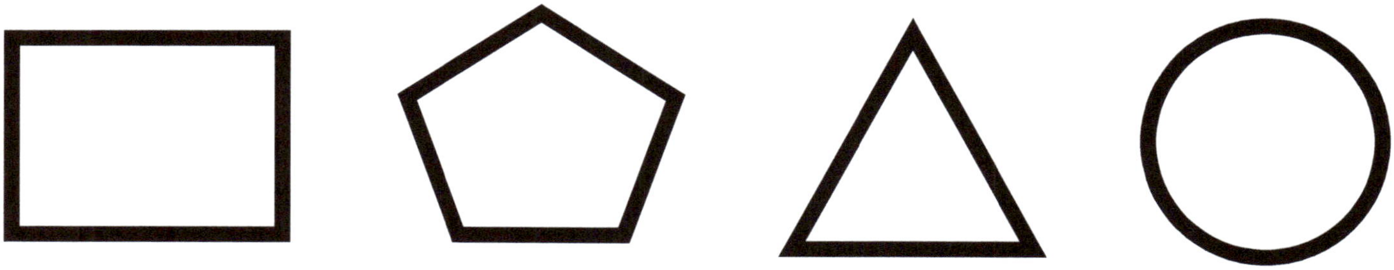

Mark the shape you see on a pyramid.

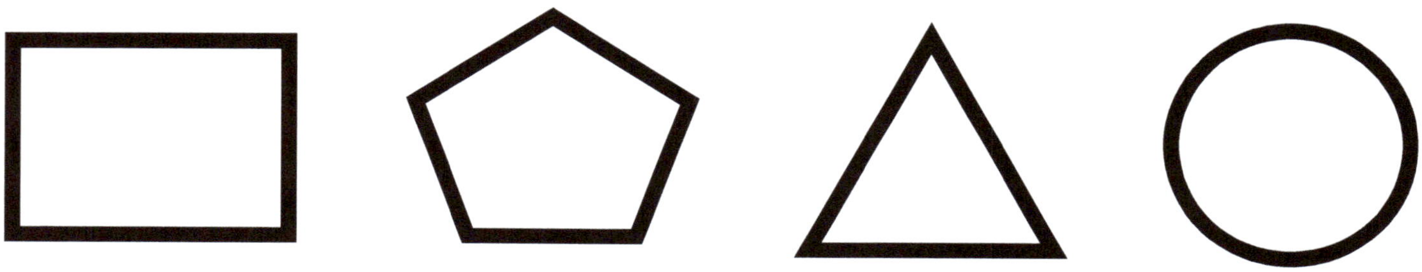

Mark the shape you see on a the end of a cylinder.

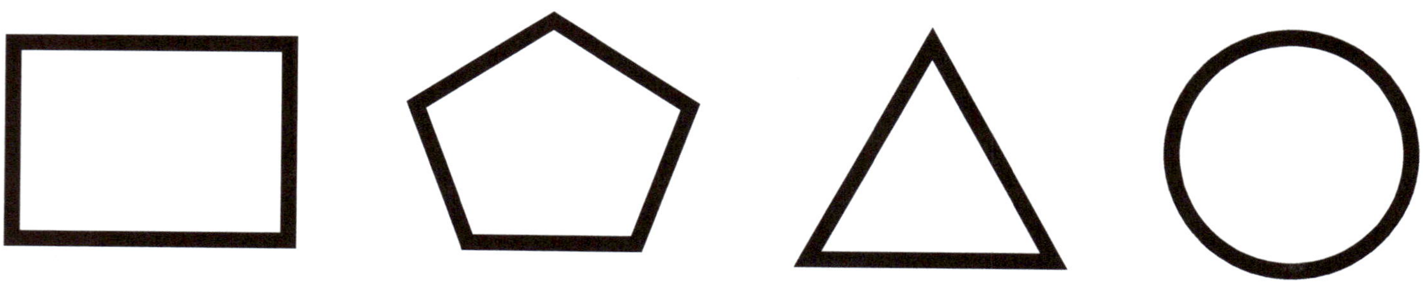

RHOMBI'S ADVENTURES IN GEOMETRY
UNDERSTAND EQUALITY

Draw a line to break each shape into two equal parts.

Draw the other half of each shape.

Glue the cut shapes to this page to make a picture.

DO NOT COPY FOR STUDENT
one-on-one with teacher

Show the child 2 pennies. Ask, "How many objects are in my hand?"

Child immediately says, "2." Has solid sense of abstract number 2...
Ready for discussions without visuals...

Child counts out loud and answers "2." Has basic understanding of abstract number 2...
Ready for handouts...

Child points to each object, counting out loud and answers "2."
Needs concrete objects to count while transitioning to abstract number sense...

Repeat actions for numbers 3 and 4.

Recognizing the sum of a pile of objects beyond 4 is almost impossible. Try it yourself. You will reflexively break the larger number into two or more lesser groups before combining to say the greater number.

All students will break numbers beyond 4 into smaller "piles." Note which students count by ones beyond 4 and which students automatically do some mental addition to give you the whole number.

RHOMBI'S ADVENTURES IN 2D
PRE AND POST ASSESSMENT PAGES

Give assessments prior to starting the projects so that you have a baseline for extension and reinforcement.

Give the same assessment after completing the projects to determine continuing areas of need.

RHOMBI'S ADVENTURES IN GEOMETRY
ASSESS NUMERACY & CARDINALITY

Print rubric for your records. Perform assessment as early as possible.

STUDENT NAME	1	2	3	4	5	6	7	8	9

NOTE: All students will need to break amounts into smaller groups in order to recognize the cardinal numbers beyond "4." The brain naturally grasps groups of 1,2,3,4. It's why phone numbers are broken into smaller groups.

RHOMBI'S ADVENTURES IN GEOMETRY
ASSESS OBJECT PERMANENCE

DO NOT COPY FOR STUDENT
one-on-one with teacher

Show the child your hand as you fold 2 pennies into your closed fist. Reopen the same hand. Ask. How many objects are in my hand?"

Child immediately says, "2." Has solid sense of object permanence for 2...
Ready for discussions without visuals...

Child counts out loud and answers "2." Has basic understanding of object permanence for 2. Ready for handouts...

Child points to each object, counting out loud and answers "2."
Needs concrete objects to count while transitioning to object permanence...

Repeat actions for numbers 3 - 9.

Remember, in this assessment, your goal is to determine a student's grasp of object permanence - NOT ability to count. The outcome is dependent on the student's ability to restate the cardinal number.

STEAMStart
COVERING NGSS

NGSS K 2		*Module*	MODULE 1					GIFTS
		STEMvestigations	ROUND	SQUARE	TRIANGLE	FIVES	MOVES	
		Engineering Design Project	Cart	Tool	Bridge	Parachute	Move	
K-2	K-2.Engineering Design							
	K-2-ETS1-1	Ask questions, make observations, and gather information about a situation people want to change to define a simple problem that can be solved through the development of a new or improved object or tool.	x	x	x	x	x	x
	K-2-ETS1-2	Develop a simple sketch, drawing, or physical model to illustrate how the shape of an object helps it functionas needed to solve a given problem.	x	x	x	x	x	x
	K-2-ETS1-3	Analyze data from tests of two objects designed to solve the same problem to compare the strengths and	x	x	x	x	x	x
K	Forces and Interactions: Pushes and Pulls							
	K-PS2-1	Plan and conduct an investigation to compare the effects of different strengths or different directions of pushes and pulls on the motion of an object.	x		x	x	x	
	K-PS2-2	Analyze data to determine if a design solution works as intended to change the speed or direction of an object with a push or a pull.*	x			x	x	
K	Interdependent Relationships in Ecosystems: Animals, Plants, and Their Environment							
	K-LS1-1	Use observations to describe patterns of what plants and animals (including humans) need to survive					x	
	K-ESS2-2	Construct an argument supported by evidence for how plants and animals (including humans) can change the environment to meet their needs					x	
	K-ESS3-1	Use a model to represent the relationship between the needs of different plants or animals (including humans)		x	x		x	
	K-ESS3-3	Communicate solutions that will reduce the impact of humans on the land, water, air, and/or other living things in the local environment.		x	x		x	
K	Weather and Climate							
	K-PS3-1	Make observations to determine the effect of sunlight on Earth's surface.		x				
	K-PS3-2	Use tools and materials to design and build a structure that will reduce the warming effect of sunlight on an area.						
	K-ESS2-1	Use and share observations of local weather conditions to describe patterns over time.	x	x		x	x	
1	Waves: Light and Sound							
	1-PS4-1	Plan and conduct investigations to provide evidence that vibrating materials can make sound and that sound can make materials vibrate.						
	1-PS4-2	Make observations to construct an evidence-based account that objects can be seen only when illuminated.	x	x				
	1-PS4-3	Plan and conduct an investigation to determine the effect of placing objects made with different materials in the path of a beam of light.		x				
	1-PS4-4	Use tools and materials to design and build a device that uses light or sound to solve the problem of communicating over a distance.	x					

RHOMBI'S ADVENTURES IN GEOMETRY
TEACHER'S PRINTABLE UNIT PLANNING PAGE

2D SHAPES: UNIT ______________________					
STEMVESTIGATION	**DATE**	**TIME**	**WHAT I'VE GOT**	**WHAT I NEED**	**PLANNING NOTES**
SCIENCE					
TECHNOLOGY					
ENGINEERING					
ARTS					
MATH					
CHALLENGE					

RHOMBI'S ADVENTURES IN GEOMETRY
TEACHER'S PRINTABLE CHALLENGE PLANNING PAGE

2D SHAPES: UNIT ______________________________

STEMVESTIGATION	**DATE**	**TIME**	**WHAT I'VE GOT**	**WHAT I NEED**	**PLANNING NOTES**
CHALLENGE:					
BRAINSTORM & HYPOTHESIZE					
DESIGN					
BUILD					
TEST					
ANALYZE DATA					

	WEEK 1	WEEK 2	WEEK 3	WEEK 4	WEEK 5	WEEK 6	WEEK 7	WEEK 8
MOON PHASES	30 MIN	CHALLENGE, BRAINSTORM, DESIGN, BUILD, TEST & USE DATA TO MAKE DESIGN CHANGES						
	60 MIN							
	45 MIN							
SHAPE WALK	30 MIN							
UNIT 2: SQUARE ONE								
GROWING THINGS			45 MIN	CHALLENGE, BRAINSTORM, DESIGN, BUILD, TEST & USE DATA TO MAKE DESIGN CHANGES				
PHOTOGRAPHY			90 MIN					
SHAPE WALK			45 MIN					
QUILT PATTERNS			60 MIN					
PAPER SQUARES			30 MIN					
GROW POLYGONS					90 MIN	CHALLENGE, BRAINSTORM, DESIGN, BUILD, TEST & USE DATA TO MAKE DESIGN CHANGES		
DIGITAL PRESENTATIONS					90 MIN			
STRUCTURES					45 MIN			
TRIANGLE ART					45 MIN			
UNIT 4: TAKE FIVE								
STAKE OUT							60 MIN	CHALLENGE, BRAINSTORM, DESIGN, BUILD, TEST & USE DATA TO MAKE DESIGN CHANGES
TOOLS							60 MIN	
MATERIALS							45 MIN	
STORY IN 5 BOXES							60 MIN	
THE PENTAGON, USA							45 MIN	

SECOND GRADE
VOCABULARY

EARTH AND SPACE	
DATA	SCIENCE INFORMATION
CONCLUSION	THE LAST PART
PREDICT	TO SAY WHAT YOU THINK WILL HAPPEN
DESCRIBE	SAYING OR DRAWING WHAT YOU SEE, HEAR, TOUCH, TASTE OR SMELL
OBSERVE	TO SEE, HEAR, TOUCH, TASTE OR SMELL
RECORD	TO WRITE OR DRAW
IDENTIFY	TO TELL WHAT SOMETHING IS
INVESTIGATE	TO GATHER INFORMATION
EVIDENCE	INFORMATION OR FACTS THAT MAKE YOU BELIEVE SOMETHING IS TRUE
ANALYZE	TO OBSERVE SOMETHING CAREFULLY IN ORDER TO UNDERSTAND IT
PHYSICAL PROPERTIES	
HAND LENSES	A HAND HELD MAGNIFYING GLASS
MASS	THE AMOUNT OF MATTER IN AN OBJECT
MATTER	ANYTHING THAT TAKES UP SPACE AND HAS MASS
NOTEBOOK	A BOOK OF BLANK PAGES FOR NOTES
OBJECT	THING
PATTERN	SHAPES, COLORS, OR LINES PUT TOGETHER IN A CERTAIN WAY
RELATIVE	COMPARING TO
TEXTURE	THE APPEARANCE AND FEEL OF A SURFACE
TOOLS	SOMETHING USEFUL
PHYSICAL PROPERTY	ATTRIBUTES THAT CAN BE OBSERVED AND MEASURED
WATER	
CONDENSATION	THE CHANGE FROM A GAS TO A LIQUID
COOLING	TO MAKE LESS WARM
ENERGY	THE ABILITY TO CAUSE MOVEMENT OR CREATE CHANGE
EVAPORATION	THE CHANGE FROM A LIQUID TO A GAS
FREEZING	TO CHANGE FROM A LIQUID TO A SOLID BY LOSING HEAT
HEAT ENERGY	THE ENERGY OF MOVING PARTICLES
HEATING	TO MAKE WARM OR HOT
MELTING	TO CHANGE FROM A SOLID TO A LIQUID BY ADDING HEAT
TEMPERATURE	HOW HOT OR COLD SOMETHING IS
THERMOMETER	A TOOL USED TO MEASURE TEMPERATURE

SECOND GRADE
VOCABULARY

MOTION	
FORCE	PUSH OR PULL
MOTION	THE ACT OF CHANGING PLACE OR POSITION
PATTERN	A REPEATING EVENT
POSITION	WHERE AN OBJECT IS PLACED
SLIDING	TO MOVE FROM ONE PLACE TO ANOTHER WITH THE SAME PART OF AN OBJECT
TOUCHING	NOT BENT OR CURVED
ROLLING	TO MOVE FROM ONE PLACE TO ANOTHER BY TURNING OVER AND OVER
SPINNING	REPEATING MOTION THAT TURNS AROUND A POINT
FAST	TO MOVE QUICKLY
SLOW	TO MOVE AT A LOW SPEED
PROPERTIES OF MATTER	
MATTER	ANYTHING THAT HAS MASS AND TAKES UP SPACE
PHYSICAL PROPERTY	ATTRIBUTES THAT CAN BE OBSERVED AND MEASURED
SHAPE	THE OUTWARD FORM OF AN OBJECT
MASS	THE AMOUNT OF MATTER IN AN OBJECT
TEMPERATURE	A MEASURE OF THE AMOUNT OF HEAT
TEXTURE	HOW A SUBSTANCE FEELS
FLEXIBILITY	ABILITY TO BEND
SOLID	A STATE OF MATTER THAT HAS A DEFINITE SHAPE AND VOLUME
LIQUID	A STATE OF MATTER THAT HAS A DEFINITE VOLUME BUT TAKES THE SHAPE OF
CAREERS	
ASTRONOMER	SCIENTIST THAT STUDIES CELESTIAL PHENOMENA
CONTRACTOR	A GENERAL CONTRACTOR IS RESPONSIBLE FOR THE DAY-TO-DAY OPERATIONS OF A CONSTRUCTION SITE. CONTRACTORS MANAGE SALES PEOPLE, BUILDERS, PLUMBERS, ELECTRICIANS, AND ALL THE OTHER TRADES REQUIRED TO COMPLETE A PROJECT. THE CONTRACTOR ALSO COMMUNICATES INFORMATION TO ALL THE PEOPLE INVOLVED IN THE PROJECT.
DESIGNER	INDUSTRIAL DESIGNERS DEVELOP CONCEPTS AND SPECIFICATIONS THROUGH COLLECTION, ANALYSIS AND SYNTHESIS OF DATA GUIDED BY THE SPECIAL REQUIREMENTS OF THE CLIENT OR MANUFACTURER. DESIGNERS MAKE CLEAR AND CONCISE RECOMMENDATIONS THROUGH DRAWINGS,VERBAL DESCRIPTIONS, AND MATH MODELS.
ENGINEER	PROFESSIONA THAT APPLIES PRINCIPLES OF SCIENCE AND MATHEMATICS BY WHICH THE PROPERTIES OF MATTER AND THE SOURCES OF ENERGY IN NATURE ARE MADE USEFUL TO PEOPLE.

VOCABULARY FOR GRADE 1
WORD WALL WORDS FOR MATH

The one who does all the work also does all the learning. Students should illustrate the words for a word wall. Just looking at the cute cards we all like to print and laminate is not particularly effective as a learning tool. While the student created cards take longer (and look significantly messier), the benefits are greater.

Addition	Making ten	Unknown
Subtraction	Doubles	Digits
Sum	Combinations	Two-digit number
Difference	Equal sign	Greater than sign
Group	True	Less than sign
Counting On	False	Mental math

Centimeter	Trapezoid	Cylinder
Inch	Half-circle	Half
Hours	Quarter-circle	Fourth
Half hour	Cube	Quarter (fraction)
Minutes	Right rectangular	Nickel
Digital	prism	Unit
Clock	Cone	

COMMON CORE
DO OR DON'T. THERE IS NO "I CAN."

A famous resident of Dagobah once said, "Do or do not. There is no try." Jedi Master Yoda may not sit with students in today's classrooms, but modern educational pedagogy would benefit from the concept. The use of "I CAN" statements is bothersome in that knowledge unshared does not contribute to community or society.

The "I can" is unnecessary. Let's help students focus on clear statements of ability. Action versus ability leaves little room for question. Goals are clear-cut, and students take pride in achievement. Visit the Ten80 Elementary wiki site to download pages that help students take charge of their own learning goals. *(available December 2014)*

I count.
I read.
I name.
I write.
I participate.
I understand.

CCSS Mathematics	Counting & Cardinality		K
Indicator		**Date**	**Date Assessed**
Know number names and the count sequence.			
CCSS.MATH.CONTENT.K.CC.A.1	I count to 100 by ones and tens.		
CCSS.MATH.CONTENT.K.CC.A.2	I count forward starting at any number I have learned.		
CCSS.MATH.CONTENT.K.CC.A.3	I write numbers from 0 to 20.		
CCSS.MATH.CONTENT.K.CC.A.3	I write a number to tell about a group of 0 to 20 things.		
Count to tell the number of objects.			
CCSS.MATH.CONTENT.K.CC.A.4	I understand how number names go with counting things in the right order.		
CCSS.MATH.CONTENT.K.CC.A.4.A	I name the number for each thing in a group as I count them.		
CCSS.MATH.CONTENT.K.CC.A.4.B	I understand that the last thing I count tells the number of things in a group.		
CCSS.MATH.CONTENT.K.CC.A.4.B	I know the number of items in a group won't change even when items move.		
CCSS.MATH.CONTENT.K.CC.A.S	I count up to 10 to tell how many things are in a group.		

CCSS Mathematics Base Ten K

Indicator		Date Taught	Date Assesse
Work with numbers 11-19 to gain foundations for place value.			
CCSS.MATH.CONTENT.I.NBT.A.1	I count up to 20 starting at any number under 20.		
CCSS.MATH.CONTENT.I.NBT.A.1	I read and write my numbers to show how many objects are in a group (up to 20).		
Understand place value.			
CCSS.MATH.CONTENT.I.NBT.B.2	I tell how many tens and how many ones are in a number.		
CCSS.MATH.CONTENT.I.NBT.B.2.A	I show that I know what a "ten" is.		
CCSS.MATH.CONTENT.I.NBT.B.2.B	I show that any number between 11and 19 is a group of ""ten"" and a certain number of ones.		
CCSS.MATH.CONTENT.I.NBT.B.2.C	I show that I understand the numbers I use when I count by tens,have a certain number of tens and 0 ones.		
CCSS.MATH.CONTENT.I.NBT.B.3	I compare two-digit numbers using <, =,and > because I understand tens and ones.		
Use place value understanding and properties of operations to add and subtract.			
CCSS.MATH.CONTENT.I.NBT.C.4	I use math strategies to help me solve and explain addition problems within 100.		
CCSS.MATH.CONTENT.I.NBT.C.4	I use objects and pictures to help me solve and explain addition problems within 100.		
CCSS.MATH.CONTENT.I.NBT.C.4	I add ones and tens when adding 2-digit numbers.		
CCSS.MATH.CONTENT.I.NBT.C.4	Sometimes I have to compose a group of higher number when adding 2-digit numbers.		
CCSS.MATH.CONTENT.I.NBT.C.S	I find 10 more or 10 less in my head.		
CCSS.MATH.CONTENT.I.NBT.C.6	I use different strategies to subtract multiples of 10 (10-90) from numbers under 100. I write the matching number sentence and explain my strategy.		

CCSS Mathematics Measurement & Data K

Indicator		Date Taught	Date Assessed
Measure lengths indirectly and by iterating lengths units.			
CCSS.MATH .CONTENT.l.MD.A.l	I put three objects in order from longest to shortest and compare their lengths.		
CCSS.MATH .CONTENT.l.MD.A.2	I talk about the length of an object using whole numbers.		
CCSS.MATH .CONTENT.l.MD.A.2	I show that I understand how to measure something by using a smaller object as a measurement tool.		
Tell and write time.			
CCSS.MATH .CONTENT.l.MD.B.3	I use any kind of clock to find the time in hours and minutes.		
Represent and interpret data.			
CCSS.MATH.CONTENT.l.MD.C.4	I organize, show and explain number information in a way that makes sense.		
CCSS.MATH .CONTENT.l.MD.C.4	I ask and answer questions about number information that is organized.		

CCSS Mathematics Geometry 1

Indicator		Date	Date Assessed
CCSS.MATH .CONTENT.l.G .A.l	I understand and explain what makes shapes unique (edges, vertices, angles as sides and corners).		
CCSS.MATH .CONTENT.l.G.A.l	I build and draw shapes that have certain parts.		
CCSS.MATH .CONTENT.l.G .A.2	I make and draw two-dimensional shapes (rectangles, squares, trapezoids, triangles, half-circles and quarter-circles).		
CCSS.MATH .CONTENT.l.G .A.2	I make and draw three-dimensional shapes (cubes, right rectangular prisms, right circular cones and right circular cylinders).		
CCSS.MATH .CONTENT.l.G.A.2	I use two-and three-dimensional shapes to create new shapes.		
CCSS.MATH .CONTENT.l.G.A.3	I understand that "halves" means two equal parts and "fourths" or "quarters" means four equal parts.		
CCSS.MATH.CONTENT.l.G.A.3	I break circles and rectangles into equal parts and use the words whole, halves, fourths, and quarters to talk about them.		
CCSS.MATH .CONTENT.l.G.A.3	I understand that breaking circles or rectangles into more equal parts means that the parts will be smaller.		

CCSS Reading	Foundations 1		
Indicator		**Date Given**	**Date Tested**
Print Concepts			
CCSS.ELA-Literacy.RF.1.1	I know how books can be read.		
CCSS.ELA-Literacy.RF.l.l.A	I find and point to the first word,the beginning capital letter and the ending mark in a sentence.		
Phonics and Word Recognition			
CCSS.ELA-Literacy.RF.1.2	I show that I know how words and their parts go together.		
CCSS.ELA-Literacy.RF.1.2.A	I explain the difference between short and long vowel sounds when I hear a word.		
CCSS.ELA-Literacy.RF.1.2.B	I put sounds together to speak words.		
CCSS.ELA-Literacy.RF.1.2.C	I find and tell the beginning,middle and ending sounds in short words.		
CCSS.ELA-Literacy.RF.1.2.D	I break down short words and say each sound by itself.		
CCSS.ELA-Literacy.RF.1.3	I show what I have learned about letters and sounds by reading words.		
CCSS.ELA-Literacy.RF.1.3.A	I hear and spell consonant letters that are blended together to make words.		
CCSS.ELA-Literacy.RF.1.3.B	I read short words.		
CCSS.ELA-Literacy.RF.1.3.C	I read words with long vowel sounds (silent e or vowel teams).		
CCSS.ELA-Literacy.RF.1.3.D	I use what I know about vowel sounds to help me figure out how many sylla-bles are in words.		
CCSSELA-Literacy.RF.1.3.E	I read words with two syllables.		
CCSS.ELA-Literacy.RF.1.3.F	I read words that have different endings on them.		
CCSS.ELA-Literacy.RF.1.3.G	I read first grade words that aren't spelled the way they sound.		
CCSS.ELA-Literacy.RF.1.4	I read and understand books at my level well.		
CCSS.ELA-Literacy.RF.1.4.A	I read and understand first grade books.		
CCSS.ELA-Literacy.RF.1.4.8	I read aloud like a teacher.		

CCSS Reading Literature 1

Indicator		Date Taught	Date Assessed
Key Ideas and Details			
CCSS.ELA-LITERACY.RL.l.l	I ask and answer questions about important details in stories.		
CCSS.ELA-LITERACY.RL.1.2	I retell a story using important details and know what an author is trying to teach.		
CCSS.ELA-LITERACY.RL.1.3	I tell the characters, setting and what happens in stories.		
Character and Structure			
CCSS.ELA-LITERACY.RL.1.4	I find words about feelings in a word or story.		
CCSS.ELA-LITERACY.RL.l.S	I know difference between fiction and nonfiction.		
CCSS.ELA-LITERACY.RL.1.6	I figure out who is telling a story at different parts in the story.		
Integration of Knowledge and Ideas			
CCSS.ELA-LITERACY.RL.1.7	I learn about characters and details using pictures in the story.		
CCSS.ELA-LITERACY.RL.1.9	I compare what happens to characters in more than one story.		
Range of Reading and Level of Text Complexity			
CCSS.ELA-LITERACY.RL.l.lO	I read and understand first grade stories and poems.		

CCSS Reading Speaking & Listening 1

Indicator		Date Taught	Date Assessed
Comprehension and Collaboration			
CCSS.ELA-LITERACY.SL..l	I have detailed conversations with my friends and teachers.		
CCSS.ELA-LITERACY.SL..ll.A	I listen and take turns when I am having conversations.		
CCSS.ELA-LITERACY.SL..ll.B	I converse by listening to other people and thinking about what to say.		
CCSS.ELA-LITERACY.SL..ll.C	I ask questions during conversations to help me understand what is being shared.		
CCSS.ELA-LITERACY.SL.1.2	I ask and answer questions about things I hear and see.		
CCSS.ELA-LITERACY.SL.1.3	I ask and answer questions of a speaker to help me understand the person better.		
Presentation of Knowledge and Ideas			
CCSS.ELA-LITERACY.SL.1.4	I use details about people, places and things to help others understand better.		
CCSS.ELA-LITERACY.SL..Sl	I draw or use other ways to share my ideas.		
CCSS.ELA-LITERACY.SL.1.6	I speak and share my ideas in complete sentences when needed.		

CCSS Reading Informational Text 1

Indicator		Date Taught	Date Tested
Key Ideas and Details			
CCSS.ELA-LITERACY.RI..l	I ask and answer questions about important details in nonfiction books.		
CCSS.ELA-LITERACY.RI.1.2	I find the main topic and important details in nonfiction books.		
CCSS.ELA-LITERACY.RI.1.3	I note that people, events or ideas are connected in nonfiction books.		
Craft and Structure			
CCSS.ELA-LITERACY.RI.1.4	I ask and answer questions to help me understand new words.		
CCSS.ELA-LITERACY.RI.l.S	I understand and use helpful parts of nonfiction books to find facts and details.		
CCSS.ELA-LITERACY.RI.1.6	I find information from pictures and words in nonfiction books.		
Integration of Knowledge and Ideas			
CCSS.ELA-LITERACY.RI.1.7	I use the pictures and words in nonfiction books to tell about main ideas.		
CCSS.ELA-LITERACY.RI.1.8	I find the reasons that an author gives to help teach about the main idea.		
CCSS.ELA-LITERACY.RI.1.9	I find facts and attributes that are the same and different in two nonfiction books that teach about the same topic.		
Range of Reading and Level of Text Complexity			
CCSS.ELA-LITERACY.RI.1.10	I read and understand first grade nonfiction books.		

CCSS Reading Writing 1

Indicator		Date Taught	Date As-sessed
CCSS.ELA-LITERACY.W.I.I	I write my opinion about a topic and give reasons for my thinking.		
CCSS.ELA-LITERACY.W.1.2	I write to teach about a topic by giving facts about it.		
CCSS.ELA-LITERACY.W.1.3	I write to tell an organized story with details.		
Production and Distribution of Writing			
CCSS.ELA-LITERACY.W.I.S	I listen to others' ideas to help add details to my stories.		
CCSS.ELA-LITERACY.W.1.6	I use a computer or tablet to publish my writing.		
Research to Build and Present Knoweldge			
CCSS.ELA-LITERACY.W.1.8	I use what I have learned to answer questions or I can find out the answers somewhere else.		
CCSS.ELA-LITERACY.W.K.8	I use what I know and have read to answer questions.		

CCSS Reading	Language K		
Indicator		**Date Taught**	**Date Assessed**
Vocabulary Acquisition and Use			
CCSS.ELA-LITERACY.l.1.4	I use strategies to figure out word meanings.		
CCSS.ELA-LITERACY.l.1.4.A	I use the whole sentence to help me figure out what other words in the sentence mean.		
CCSS.ELA-LITERACY.l.1.4.B	I use the beginnings and endings of words to help me figure out what it means.		
CCSS.ELA-LITERACY.l.1.4.C	I find root words with lots of different endings.		
CCSS.ELA-LITERACY.l.l.S	I look for ways that words are related.		
CCSS.ELA-LITERACY.l.l.S.A	I sort things into groups and use the names of the groups to help me understand them better (colors, clothing).		
CCSS.ELA-LITERACY.l.l.S. B	I explain a word by telling how it belongs in a group.		
CCSS.ELA-LITERACY.l.l.S.C	I know how words are used in real-life.		
CCSS.ELA-LITERACY.l.l.S.D	I tell or show the difference between verbs (action words) that a seem alike (look, peek, glance, stare, glare, scowl).		
CCSS.ELA-LITERACY.l.l.S.D	I tell or show the difference between words that look alike.		
CCSS.ELA-LITERACY.l.1.6	I use words in different ways to show that I know what they mean.		

CCSS Reading	Language 1		
Indicator		**Date Taught**	**Date As-sessed**
Conventions of Standard English			
CCSS.ELA-LITERACY.l.l.l	I know how to use words correctly when I write and speak.		
CCSS.ELA-LITERACY.l.l.l.A	I print all of the upper and lower case letters.		
CCSS.ELA-LITERACY.l.l.l.B	I use common, proper and possessive nouns.		
CCSS.ELA-LITERACY.ll.l.C	I use singular and plural nouns with matching verbs in my sentences.		
CCSS.ELA-LITERACY.l.l.l.D	I use pronouns like I, me, my, they, them, their, anyone, and everything.		
CCSS.ELA-LITERACY.l.l.l.E	I use verbs in the right way to tell about the past, the present and the future.		
CCSS.ELA-LITERACY.l.l.l. F	I use adjectives.		
CCSS.E LA- LITE RACY.l.l.l.G	I use conjunctions.		
CCSS.ELA-LITERACY.l.l.l.H	I use determiners.		
CCSS.ELA-LITERACY.l.l.l. I	I use common prepositions like during,beyond,and toward.		
CCSS.ELA-LITERACY.l.l.l.J	I use simple and compound sentences.		
CCSS.ELA-LITERACY.l.1.2	I know how to write sentences correctly.		
CCSS.ELA-LITERACY.l.1.2.A	I use capital letters in dates and also when I write people's names.		
CCSS.ELA-LITERACY.l.1.2.B	I use the right punctuation at the ends of my sentences.		
1CCSS.ELA-LITERACY.l.1.2.D	I use sight words and spelling patterns to help spell words correctly.		
1CCSS.ELA-LITERACY.l.1.2.E	I use what I know about letters and their sounds to spell new words correctly.		
CCSS.ELA-LITERACY.L.K.4.B	I use beginnings and endings of words to help me figure out what they mean.		
CCSS.ELA-LITERACY.L.K.S	I figure out how words are related. I figure out how meanings might be alike.		
CCSS.ELA-LITERACY.L.K.S.A	I sort things into groups. I use groups to understand things better.		
CCSS.ELA-LITERACY.L.K.S.B	I match some action verbs and adjectives with their opposites.		
CCSS.ELA-LITERACY.L.K.S.C	I tell how words are used in real-life.		
CCSS.ELA-LITERACY.L.K.S.O	I tell the difference between verbs (action words) that are almost alike.		
CCSS.ELA-LITERACY.L.K.6	I use new words in different ways to show that I know what they mean.		

RHOMBI'S ADVENTURES IN 2D
MATH MODELING IN K-2 SETTINGS

DEFINING THE MATH MODEL

Mathematics can be used to "model", or represent, how the real world works. A model is not the real thing, but it should be accurate enough to be useful. Math models connect abstract numbers to real experience. Math models save time, supplies, and aggravation. For example, Rhombi needs to build a full-scale cart. Since the class cannot build the full sized vehicle, it will develop a small scale version and extrapolate from there. They use their background knowledge to begin the math model.

MOVE FROM EXPERIENCE TO SYMBOLS

The study of science and math started in an effort to understand actions that could be observed directly. In order to make science and math a useful performing art, instruction must start with real experience and gradually build toward an abstract representation. Doing things in the reverse chronological order produces nearly useless results.

KNOW WHERE THE LESSON IS HEADED

Education separates experience from the learning process. We could learn a few things from our own history. Math development progressed along its slow, continuous track: caveman to calculus. Here, in the 21st century, we expect second graders to operate at abstract levels using a symbolic language that took mathematicians thousands of years to develop. Educators are asked to start with the algorithm –the abstraction- and work backwards to the concrete. We wonder why math is such a dreaded subject. Neither the teacher nor the student knows where the symbols are supposed to take them.

STEAM USES NATURAL EXTENSIONS

Abstract models of STEAM phenomena should grow as a natural extension of what is felt by direct experience. The abstraction must be an immediate and obvious representation of what students already know at every level. Drive a radio controlled car. Build the shapes as you memorize them. Investigate, collect data and analyze results. From reality to abstraction, STEAM learning needs more teachers who understand that doing more means going further, faster.

DEFINE YOUR MATHEMATICAL MODEL

Identify the problem, define the terms in your problem, & draw diagrams where appropriate. Begin with a simple model, stating assumptions that you make as you focus on particular aspects of the phenomenon. Identify important variables and constants. Determine how they relate to each other. Develop the equation(s) that express relationships between variables & constants.

RHOMBI'S ADVENTURES IN 2D
REFERENCE MATERIALS AND SUGGESTED READING

"Circle." Wikipedia, the Free Encyclopedia. Wikimedia Foundation, Inc, n.d. Web. 7 July 2014.

Dehaene, Stanislas. *Number Sense: How the Mind Creates Mathematics.* New York: Oxford UP, 1997. Print.

Devlin, Keith J. *The Math Gene: How Mathematical Thinking Evolved and Why Numbers Are Like Gossip.* New York: Basic Books, 2000. Print.

Ferris Bueller's Day Off. Dir. John Hughes. Perf. Matthew Broderick, Mia Sara, Alan Ruck, Jennifer Grey, and Jeffrey Jones. Paramount, 1986. Film.

Feynman, Richard P, and Jeffrey Robbins. *The Pleasure of Finding Things Out: The Best Short Works of Richard P. Feynman.* Cambridge: Perseus Books,1999. Print.

Furoy, Michael, Rosemary T. Wong, and Harry K. Wong. *The Effective Teacher.* Mountain View: Harry K. Wong Publications, Inc., 2009. Print.

Ifrah, Georges. *From One to Zero: A Universal History of Numbers.* New York:Viking, 1985. Print.

Lee, Joon S., and Herbert P. Ginsburg. "Preschool Teachers' Beliefs About Appropriate Early Literacy and Mathematics Education for Low Socio-Economic and Middle-Socioeconomic Status Children." Early Education and Development 18.1 (2007): 111-143. Print.

Ma, Liping. *Knowing and Teaching Elementary Mathematics: Teachers' Understanding of Fundamental Mathematics in China and the United States.* Hillsdale, NJ: Lawrence Erlbaum Associates, 1999. Print.

Mathemagic. Chicago: World Book—Childcraft International, Inc., 1980.

McLeish, John. *The Story of Numbers.* New York: Fawcett Columbine, 1994.

Next Generation Science Standards. N.p., n.d. Web. 7 July 2014.

Wassily Kandinsky – biography, paintings, books. N.p., n.d. Web. 7 July 2014.

"Inquiry Based Science: What Does It Look Like?" Connect Magazine. 1995:13.

www.ingramcontent.com/pod-product-compliance
Lightning Source LLC
LaVergne TN
LVHW071345210926
848486LV00003B/235